Field Expedient SDR
Volume 1

Introduction to Software Defined Radio

Paul Clark

David Clark

Meadow Registry Press

Field Expedient SDR: Introduction to Software Defined Radio

To:

MQ and LVP

Y and S and K

and our parents, who invested in love and education

Preface

Intended Audience

The word fun comes from the word fundamentals. We believe that learning how something works (for instance a car, computer, or radio) enables a person to better interact with that something. Learn the fundamentals of a topic, and then leverage that to have fun.

The Internet is useful for looking up information on just about any topic, but sometimes it can take a while to assemble data from a vast array of different web sites into one cohesive mass of knowledge. This book will not eliminate the need for research, but we hope it will provide some of the fundamentals to better grasp the interesting topic of Software Defined Radio (SDR).

We have had people tell us that on the Internet they could find software for running SDR hardware, but it was not clear how the various components worked. There was either "too much math" to describe various SDR functions or too little information.

Rather than develop your understanding of radio theory through formal mathematical means, this book adopts a hands-on approach. Each project you work through will advance your understanding of a key radio concept and show you how to implement it using SDR.

This book is written both for those who have no SDR experience as well as those who have struggled to get off the ground with SDR. You may be a tinkerer, an amateur radio enthusiast, or a student. Alternatively, you may be an engineer who's forgotten most of the radio theory you learned in school and want a refresher. We just ask for you to bring your curiosity to this subject. We warn any experts in the field of radio or digital signal processing that they may be bored and/or insulted by this book, and that we will not pay for therapy required due to reading this material.

It is useful if you have some Linux command line knowledge. There are many resources on the Internet to acquire the basics of the command line. There is also a book entitled *Field Expedient Linux* that may be helpful.

General Layout of This Book Series

Volume 1 - Field Expedient SDR: Introduction to Software Defined Radio explains complex but foundational radio concepts through a series of projects that are easy to build and understand.

Volume 2 - Field Expedient SDR: Basic Analog Radio dives deeper into SDR theory and operation while also focusing on the intensely practical skills needed to make real-world radios work.

Volume 3 - Field Expedient SDR: Basic Digital Communications covers the techniques used to transmit and receive digital data via radio.

Volume 4 - Field Expedient SDR: Introduction to Reverse Engineering explores the art and science of analyzing unknown signals and then breaking them down to reveal the data within.

Where to Find the Files for This Book

To make your SDR learning experience as smooth as possible, we've provided a number of project and input data files on our web site at *www.fieldxp.com*. We recommend that you uncompress the contents into a convenient location on your hard drive, because we will be using the files frequently.

The contents of the compressed file are broken down by chapter, so you can easily find the files you need as you work through the material. We recommend that you build each project from scratch as shown in the book's text, but we included the finished projects as a reference in case you run into trouble.

Each chapter will typically contain one or more input data files in addition to completed projects. Thanks to these input data files, it is possible to work through much of the material in this book without needing access to any SDR hardware.

Recommended Hardware

Volume 1 for this series does not require any SDR hardware. We recommend the HackRF One SDR hardware as a good starting point for those desiring to explore this subject matter.

A Note about Fonts

You'll also notice that we use a number of different fonts and text formats throughout the book. When you encounter the bold Courier font, this indicates text that is typed directly into a Linux terminal or text editor, such as:

`mkdir sdr`

or

change the value of frequency to `880e3`

The Courier font is also used to highlight items selected from user interface menus:

for the Type property we select `Float`

We also reference the names of files that you can find in the project folder associated with this course. We have underlined those files names in the text, as in the following example:

... after you open the file simple_multiply.grc

Miscellaneous

This book relies heavily on screenshots to illustrate the various projects we cover. The images in the book will also look slightly different from what appears on your screen because you may be using a different version of gnuradio and also because we enlarged certain gnuradio display features for printing.

This book is written in a conversational tone, often including phrases like "Remember when I told you..." Although this book has two authors you will see that we use the term "I" rather than "we" throughout the book. We believe this choice makes the text more readable.

Table of Contents

1 *Introduction*

1.1 What are These Things?

Since Marconi's first transmissions over 100 years ago, nearly all radios have been fixed-function devices. By that I mean that you could only tune them to a relatively narrow range of frequencies and transmit or receive a specific kind of signal. At most, you'd have combination devices – like those that received both AM and FM signals. Still, these were largely just two (or more) fixed-function radios combined.

Another way to describe these common radios is *Hardware Defined Radios*. What defines the frequencies these radios can transmit or receive? Hardware. What defines if it's an FM radio or a Wifi router? Hardware. Nearly every aspect of these radios is defined by stubbornly inflexible hardware.

On the other hand, Software Defined Radios (SDRs) provide programmability throughout the radio's architecture. When you want to change the operation of an SDR, you don't rewire your hardware, you simply modify the programming on the SDR board (and quite often on a connected host computer).

1.2 What Can I Do with Them?

It may not be immediately obvious how powerful an SDR can be, so let me run through a few scenarios.

It's Monday and you want something to talk to a number of WiFi devices. You put together a project on your computer, connect to your SDR and sure enough - you have a WiFi access point. And not just any access point, but one in which you have control over all aspects of the channels used and the packets sent.

On Tuesday you need some geographical data from a job site you are visiting. With nothing more than some software changes you reprogram your SDR to be a GPS receiver. As before, you're not just limited to basic GPS functionality, but can keep track of how many satellites in the GPS constellation are observable and the signal strength of each. On the way back from the job site, you program your SDR to pick up FM radio stations so you can listen to a radio show.

Sometime on Wednesday you realize that your WiFi system is having some unknown difficulties. Suspecting security issues, you reconfigure your SDR to scan all 14 available WiFi channels simultaneously and see what kind of traffic is out there. You also save all the raw RF data to a file for further processing at your leisure.

You shift gears on Thursday, turning your attention to some signals you saw broadcasting in the 2.4 GHz WiFi band the day before. Not sure what it is, you start breaking down the signals using powerful but free and easy-to-use software. The mystery signals are not currently transmitting, but that's not a problem, because you have the raw data from yesterday. After looking at the data, you're able to determine that there's a ZigBee home automation network nearby, as well as a baby monitor and a poorly shielded microwave oven.

Now comes Friday, and you decide to drive around and check out the RF activity at various points around your city. Taking advantage your WiFi and GPS routines from earlier in the week, you're able to write a position-aware data-logging application in just a couple dozen lines of Python code.

For the weekend camping trip, you reprogram your SDR to several different amateur radio modes in hopes of making contact with other Ham operators. Perhaps another connection may be made with the International Space Station.

Are you starting to see how powerful these things are? Not only can you easily implement new radio designs, but you can also switch between them with nothing more than a few keystrokes or a function call in your code. And more than simply transmitting and receiving, you can scan, find, and deconstruct other signals that may be out there.

In the interests of full disclosure, let me mention one caveat to all this. As programmable as SDRs have become, their creators still haven't figured out how to make affordable programmable antennas. So for now, you may still have to swap those out when making big frequency changes.

1.3 The Bigger Picture

There is another key application for SDRs – prototyping new products containing an RF component. In the early stages of product development, it is critical to iterate quickly on new designs. Fail fast and try as many new ideas as reasonably possible.

SDRs allow you to implement the radios in your design far more quickly than designing them from off-the-shelf components. You can also modify their functionality far more quickly. Due to cost constraints, you most likely won't go to production with an SDR in a high-volume product, but you can optimize your product for cost later in the design cycle.

We've seen this before. For decades, digital logic was typically a hardwired affair. You grab a bunch of chips containing logic gates, wire them together and voila! – there's your system. If performance or cost demanded, you could even have a custom chip made (they are called Application Specific Integrated Circuits, or ASICs). Then a new technology came and changed that. Engineers started using a number of different types of programmable logic for prototyping, going by questionably helpful abbreviations such as PLAs, PALs and PLDs. The most commonly used form today is the Field Programmable Gate Array (FPGA).

An interesting development occurred along the way. In certain cases, engineers started realizing that the cost of the programmable solution wasn't so much higher than the fixed-function solution after all. In fact, when factoring in the engineering hours required for designing the fixed function implementation, sometimes the programmable solution was cheaper. And there was the added benefit of being able to update the product's hardware functionality at any point in the production process and beyond - even when the product was in the hands of customers.

Although I have focused our little history lesson on digital logic, similar programmable technologies now exist for analog circuitry. One could even consider 3D printers to be in the same vein – *Software Defined Matter*.

All of these technologies reduce design cycle times and get your designs out more quickly. In the early stages these designs are merely prototypes, but as the technology matures, it increasingly finds its way into released products. This happens most quickly in two circumstances. First, when time-to-market is critical, the additional cost may be acceptable. Second, when you're not making a large quantity of your product, it doesn't make sense to spend the engineering hours to optimize its cost.

How long will it take for SDRs to become ubiquitous in the marketplace? I can't tell you for sure, but you might want to do an Internet search on "RTL-SDR." In the past, this particular SDR design found its way into a number of consumer products (mostly digital TV tuners), while radio hobbyists hacked the products to turn them into extraordinarily low-cost devices for their own experimentation.

1.4 How This Book is Structured

The goal of this book is to learn by doing. This will not start with a ton of pages of dry exposition on electromagnetic theory and the mathematical underpinnings of signal processing. I will be covering these concepts, but in a very different way than you would encounter them in an academic textbook. I will start with the simplest possible concepts, pair them with actual SDR projects, and then gradually build on what we've learned with progressively deeper concepts and projects.

Imagine an onion. Only this is a software defined radio onion. In each chapter of this book, I'll be peeling back a very thin layer of that onion (hopefully without the tears). Don't worry if you don't fully grasp a topic in the early chapters, the concepts should become clearer as we work through more exercises. After a number of chapters, I think you'll be pleasantly surprised at how deeply you've traveled into the heart of the onion.

There's one catch, however - to learn by doing does require a platform for the "doing." Now, spending a significant amount of time installing software tools is annoying, but it is key to the "learn by doing" approach I am taking in this book. What this means for you is an hour or two slogging through tool installation so you can actively start playing with radios, both simulated and real. Don't worry, though, most of this time will not require your active intervention but will consist of simply waiting for installation processes to complete.

What do I mean by "simulated" radios? Well, it turns out that our primary SDR software - gnuradio - contains the ability not only to control your SDR hardware, but also to simulate the operation of a real radio without the need to hook up any hardware. This simulation capability is the key to how we're first going to learn to build software defined radios. For each concept or radio design that I cover, I will first go through a simulated project before moving on to a hardware-based project. Because you can do so much with the simulator (and the input files provide by the authors), you can actually learn quite a bit about software defined radios without purchasing a single bit of additional hardware.

Which brings us to my final point. If you can't get your hands on any SDR hardware, that doesn't need to stop your radio learning. All you need is your computer and some open-source software and you can do far more than you've probably imagined.

1.5 Recommended Reading

I anticipate different types of readers for this book, and have made an effort to make it useful and comprehensible to a wide array of backgrounds. Because this book begins with the very basics and builds up through successive chapters, it may bore those who have extensive education/experience in programming or RF engineering.

If you are already familiar with radio technology and digital signal processing, you may want to skim through (or even skip if you're feeling adventurous) certain chapters in this book. However, I would suggest that you go through the exercises contained in each of the chapters. Even though you may not learn anything new about the radio theory, the lab exercises will help you better understand gnuradio.

1.6 Things Covered and Not Covered

The goal in this book series is to help you learn to do the kinds of things I talked about during our fictional "Radio Week" mentioned earlier in this chapter. For that, I strongly recommend that you eventually acquire SDR hardware and actually build the physical projects I describe throughout the book. But I hope you won't wait for that mail order SDR device to arrive. As stated previously, you can start right now.

So far, I have mentioned a few terms and concepts that may be a bit obscure. I've mentioned a few more (frequency, for example) that are probably familiar to you, but only in a general sense. My intent is to provide you with a functional understanding of the terms and concepts required to build practical SDR systems.

However, this is not a digital signal processing (DSP) textbook. I don't believe many beginners to SDR are well served by a deep dive into the detailed mathematics underpinning signal processing, sampling theory, or the study of electromagnetic fields and waves. As such, I will provide an overview of key mathematical concepts when necessary and refer you to other sources whenever I can. And I will always endeavor to not just tell you the concept, but show you the concept by actively using gnuradio.

I take the same approach with respect to computing topics. I assume that you have a basic understanding of how to work in your chosen operating system. Though you can work in OS X or Windows if you like, you will see later that I recommend Linux for your SDR work. If you're new to Linux, you might want to look through some resources online to familiarize yourself with it first. You can also refer to another book - *Field Expedient Linux* - which covers command line topics with examples.

2 *Installing the Software*

2.1 Installation

Remember the hour or two of slogging through software installation I promised (or warned) you about in the introduction? Well, it's here. Unless you happen to be a hardcore Linux guru, this will likely be your least favorite part of the book. However, if you follow the instructions carefully, you should be through the installation wilderness before you know it.

First, you need to be aware of the basic software tools out there for SDRs. We'll be working primarily in gnuradio, so installing this is a must.

Before I proceed, a word about operating systems. Two words, really.

"Use Linux."

It is definitely possible to install gnuradio on OS X and Windows. I've done it and made them work. It was not fun nor was it easy, and with every operating system update, for every new version of gnuradio released, something may change just enough to complicate things anew. For this reason I strongly advise you to follow the Linux path. The gnuradio tool has been developed natively for Linux and has always been easiest to get up and running on that platform. This will likely always be the case.

One final note: you might be tempted to create a virtual machine (VM) and install your software there. You may be able to get this to work for your software, but the hardware will likely be a different story. VMs typically have their biggest issues when dealing with hardware connectivity and performance. For this reason, I have not included any information on VM installations in this book, and will assume that you will dedicate a partition to Linux.

2.2 Binaries vs Source Code

I will use the Ubuntu Linux distribution in this book. Honestly, it is just easier for me to support one Linux distribution rather than write installation instructions for multiple Linux distributions. By the time you read this, the instructions may have changed, so see *www.fieldxp.com* for updates.

The instructions here are for Ubuntu 14.04.4. Your mileage may vary with other Ubuntu versions. One of the reasons I picked version 14.04 is because it is a Long Term Support (LTS) Ubuntu version. These instructions should also work with Ubuntu 16.04 (also an LTS version).

Some software comes pre-installed with a Linux distribution. These program are sometimes called "binaries." Additional software can be installed via the Ubuntu Software Center or Synaptic Package Manager. Many people prefer to install software using **apt-get** on the command line. The **apt-get** command is not covered here.

Yet there may be times when software must be installed from "source." This means obtaining the source code for a software application, and compiling the application from the source code.

"Source code" is the lines of computer code a software developer writes in a particular computer language such as C, C++, Java, etc.

"Compiling" is the act of taking the source code and creating a program (also known as a binary).

2.3 Why Compile from Source Code?

There are different reasons why one may want to compile a program from source instead of using the binary available from the Ubuntu Software Center or Synaptic Package Manager:

Availability: The program may not be available via the Ubuntu Software Center or Synaptic Package Manager. Some companies and software developers only make their program available via source code.

Latest version: The latest version of the software may not be the one accessible by the Ubuntu Software Center or Synaptic Package Manager. If you want the latest version then you have to compile it manually.

Features: Some software features must be enabled when the software is compiled. A binary may have certain features disabled. If the binary that someone is using lacks a feature (and the feature exists), then the software will have to be compiled with the feature enabled.

Optimization: Compiling a program allows the user to obtain better performance by targeting the machine (e.g. Intel or AMD processor) that will be running the program.

We will be using a program called PyBOMBS to install the SDR tools from source. This will enable us to run a later version of these tools than what may be available otherwise.

2.4 Steps for Installing from Source Code

While the exact steps may vary, in general one must:

1: Download (or obtain) the source code

2: Unpack the source code

3: Configure the source code

4: Build (or compile) the program binary

5: Install the program binary

6: (Optional) Uninstall the program

Fortunately, using the PyBOMBS program, much of the above will be automated. The following sections 2.5 through 2.7 are provided here as a support in case you need to install from source some software in the future. If you don't have an urgent need to install some software from source, you might want to skip over this material and go directly to section 2.8.

2.5 Obtain and Uncompress the Source Code

However the software is obtained (from a web site, CD, email attachment, etc.), it will most likely come as a compressed tarball.

A tarball is a file containing one or more files. If there are directories present, then the hierarchy of directories and files is preserved. An archive is not a compressed file, but rather an orderly collection of files.

- The ".tar" extension denotes a tarball.

The tarball is then compressed in order to reduce the size of the file. The ".gz" or ".bz" or ".bz2" extension denotes a compressed file.

- Files with the ".gz" extension are compressed with the GZip algorithm
- Files with the ".bz" extension are compressed with the BZip or BZip2 algorithm.
- Files with the ".bz2" extension are compressed with the BZip2 algorithm.

Example: After obtaining the file example.tar.gz, one must first uncompress the file.

```
gunzip example.tar.gz
```

This will expand the compressed tarball and a file named "example.tar" will be created.

2.6 Extract (un-tar) and Configure the Source Code

To unpack the tarball:
```
tar -xvf example.tar
```

- The "x" option extracts the contents of a tar file
- The "v" option verbosely lists the files inside the tar file
- The "f" option is used when specifying a file name

There will now be a directory named "example" in the current directory. Change into this directory by:
```
cd example
```

The source code is present in the "example" directory. It must now be configured before compiling. Multiple things will happen with configuring. The configure program will check the computer for items such as memory, CPU type, what tools for compiling are present, etc.

To start the configuration program, type:
```
./configure
```

- The "./" tells Linux to run the program even if it is not in the path.
- The path specifies a set of directories which the operating system will search for when the user issues a command.

2.7 Compiling and Installing the Software

The information gathered from the configure program will then be used to generate something called a makefile.

The makefile will be used by the Make program to create the program (binary) from the source code.

To create the program, type:
```
make
```

The make program will consult the makefile that specifies the order that the different source code files are to be compiled.

Alternatively, one may use the command:
```
make clean
```

- Running **make clean** can get rid of files that are not needed after compiling (and save disk space).

The time it takes to run the **make** command will depend on a number of factors such as the size and complexity of the program being compiled, the CPU and speed of the computer, etc.

- In the Linux world, the make program usually uses the "gcc" compiler to compile the software.

To now install the compiled program, type:
```
make install
```

To uninstall the program, make sure the following command is executed from the directory where the **make** command was run:
```
make uninstall
```

2.8 Installing gnuradio on Linux

So you listened to my warnings and decided to go with Linux. Great! This means, however, that you'll need a computer with Linux installed on it, specifically Ubuntu version 14.04.04 or Ubuntu version 16.04. If you don't have one of these operating systems installed and are unsure of how to do so, please find a guide online to walk you through the process. It's also a good idea to apply any Ubuntu system updates and to reboot before following the instructions in this chapter.

As you set up your Linux partitions, ensure that your computer has at least 50 GB of space available for the operating system, gnuradio, and the data files with which we'll be working. 100 GB or more would be better.

Once you have an Ubuntu installed, the general flow we're going to follow is this:

- First, we will install an application called git, which is a very common tool used to fetch files from a software repository (often online).
- Second, we will install some dependencies (software upon which other software depends).
- Next, we will download something called pip. This is a program that helps us install other programs that are written in the Python programming language.
- After that, we will use the pip tool to download and install an application called pyBOMBS. This is a special software manager that is used to install gnuradio as well as other SDR related software utilities (for instance, software that enables HackRF functionality from inside gnuradio).
- Lastly, we will handle the environment variables required by gnuradio. These are just settings that Linux needs to run gnuradio properly.

Don't worry if a few of the steps I just mentioned are a bit confusing. I'll walk you through each of the steps below, giving you the specific commands to type.

First, a word of warning. Some gnuradio install guides will direct you to run the following simple command WHICH YOU DO NOT WANT TO TYPE. But this is a command which you may see:

```
sudo apt-get install gnuradio
```

This will indeed install gnuradio, but it will not be a very recent version. The steps we'll go through below will get the latest version of gnuradio available, and this will make a difference. Several of the exercises in this book series will not work with the older version of gnuradio.

I have also provided a rough time estimate for how long each command will take. This estimate assumes a mid-range 2015 workstation and a 10 Mbit/second Internet connection (your mileage may vary).

Let us assume that we're starting in the home directory. To make sure, go ahead and type the following in a terminal window:

```
cd
```
(Time required: instantaneous)

Time to make sure Ubuntu Linux is up-to-date:

```
sudo apt-get update
```
(Time required: could take a few minutes)

First we'll install git, using apt-get. You'll need to enter your password before the command will complete.

```
sudo apt-get install git
```
(Time required: less than a minute)

Now it is time to install some dependencies, software that the SDR tools will depend on in order to install and/or run:

```
sudo apt-get install libyaml-dev
sudo apt-get install libssl-dev
sudo apt-get install python-dev
```
(Time required: less than a minute)

Next, we'll create a directory in which to place gnuradio and other software utilities. If you don't want to call it "sdr" feel free to give it another name or location, just be careful to use that other name throughout the guide in place of "sdr" name.

```
mkdir sdr
cd sdr
```
(Time required: instantaneous)

Now we'll use install a tool that will be used to manage Python packages:

```
sudo apt-get install python-pip
```
(Time required: less than 1 minute)

The next step is to update the python-pip software we just installed:

```
sudo easy_install pip
```
(Time required: less than 1 minute)

Now we use the pip tool to install PyBOMBS:

```
sudo pip install PyBOMBS
```
(Time required: less than 1 minute)

The following commands will add "recipes" for PyBOMBS. Different pieces of software will be installed by PyBOMBS via various recipes.

```
pybombs recipes add gr-recipes git+https://github.com/
gnuradio/gr-recipes.git
pybombs recipes add gr-etcetera git+https://github.com/
gnuradio/gr-etcetera.git
```

(Time required: about 1 minute)

To list the current installed recipes:

```
pybombs recipes list
```

(Time required: about 1 minute)

Next we will designate the previously created "sdr" directory as the target for software installed by PyBOMBS:

```
pybombs prefix init ~/sdr -a myprefix
```

(Time required: instantaneous)

Finally, we will use PyBOMBS to install gnuradio and gr-osmosdr (this provides gnuradio with an interface to the HackRF hardware):

```
pybombs install gnuradio gr-osmosdr
```

(Time required: hard to say, maybe 40 to 70 minutes)

With Linux, we can use the source command on a file to execute a list of commands in that file (rather than typing each command one at a time). The install process we just completed will generate a number of environment variables that need to be set for gnuradio to work. Fortunately, all of these variables have been automatically added to a single file that you can simply source with the following command:

```
source ~/sdr/setup_env.sh
```

(Time required: instantaneous)

Running this source command will be enough to get gnuradio running for now, but it will need to be run again every time you reboot your machine. To run this command automatically at boot time, edit the file called ".bashrc" in your home directory and add the source command shown above to the end of the file.

At this point, we should have a working installation of gnuradio. Go ahead and try to run its graphical interface by typing:

`gnuradio-companion`

(Time required: instantaneous)

If successful, you should see a window similar to this appear:

Both gnuradio and PyBOMBS undergo changes over time. This software work is undertaken by talented programmers in the open source community. Despite their best efforts, there may be hiccups during the installation process. The installation instructions given here may not work in the future. See the *www.fieldxp.com* website for more up-to-date instructions.

Almost done, now let's close the gnuradio window and go back to our terminal window. We need to install some additional graphical support software in gnuradio. This will enable more advanced displays so we can better see what's happening on the screen. The following will use pip to install OpenGL support for Python.

`sudo pip install PyOpenGL PyOpenGL_accelerate`

We may need to type in a few "Y" characters to make this go.

Finally, we will enable our new graphical functionality in gnuradio by creating a file using gedit (or your favorite text editor) in the following location:

```
gedit ~/.gnuradio/config.conf
```

This file needs to contain the following contents, so we'll type (or paste) it in.

```
[wxgui]
style=gl
fft_rate=30
waterfall_rate=30
scope_rate=30
number_rate=5
const_rate=5
const_size=2048
```

Now we should have everything we need to get started, including gnuradio and all of the hardware drivers and utilities needed by the HackRF unit (the SDR hardware we mentioned in the Preface). Keep in mind we will not need SDR hardware such as the HackRF for this volume.

The following is for those who do have a HackRF SDR unit. We've done a sanity check already on gnuradio, but we can also do a simple test of the HackRF utilities and drivers (think of a driver as the software that the operating system uses to manage hardware). First, we need to plug in the HackRF device to a USB port on our computer. Then we run the following command:

```
hackrf_info
```
(Time required: instantaneous)

If things are working, you should see a message telling you that a HackRF board was found, along with information about the board. That message will look something like this:

```
Found HackRF board.
Board ID Number: 2 (HackRF One)
Firmware Version: <a series of interesting characters>
Part ID Number: <more interesting characters>
Serial Number: <even more interesting characters>
```

If the hackrf_info command produces an error message, than:

```
cd ~/sdr/src/hackrf/host/build
cmake ../ -DINSTALL_UDEV_RULES=ON
make
sudo make install
sudo ldconfig
```
Now unplug the HackRF unit, plug it in again, and try the hackrf_info command.

2.9 Validating your gnuradio Installation

Regardless of which operating system you used to install gnuradio, you should create the following project just to make sure you have things working OK. This is not a thorough or exhaustive test, just a sanity check to make sure the big pieces are in place and functioning correctly.

I'll guide you through a couple of steps that may not make a lot of sense at first, but don't worry about that yet. It will become progressively more clear as I peel back the layers of the SDR onion. Please note that screenshots in the book may not match up exactly with your gnuradio-companion display (due to, for instance, different versions of gnuradio).

1) open gnuradio-companion

- For Linux, you will simply type **gnuradio-companion** in a terminal window

- When you're done, you should see something like this:

2) Type Control-F to bring up a search bar on the top right side of the screen

3) Type **signal** into the window (you don't have to hit the return key). You'll see an option under Waveform Generators called **Signal Source**. Double click it and it will appear in the diagram.

4) Erase the existing text in the search window and type **audio** in its place. You'll see an option called **Audio Sink**. Double click this and it too will appear in the diagram.

5) Double click the Signal Source block and another window will come up. The contents won't make a lot of sense, but that's OK. We only need to change two things. Enter **0.1** for the Amplitude and change the Output Type from Complex to **Float**. Then click OK.

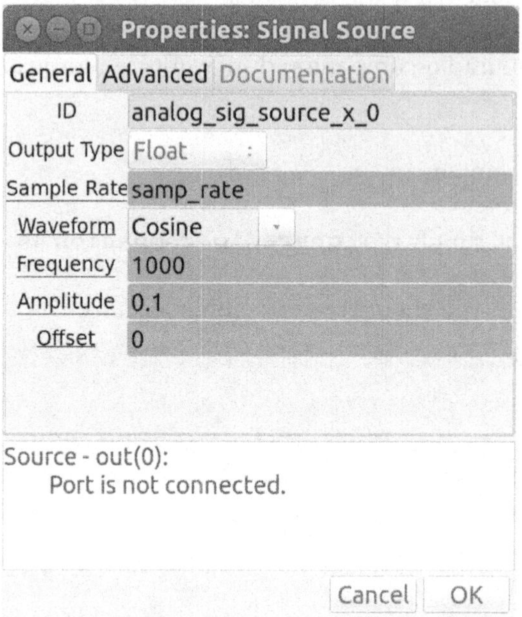

Properties: Signal Source	
General Advanced Documentation	
ID	analog_sig_source_x_0
Output Type	Float
Sample Rate	samp_rate
Waveform	Cosine
Frequency	1000
Amplitude	0.1
Offset	0

Source - out(0):
    Port is not connected.

Cancel   OK

6) Now click the orange tab on the signal source block and then click again on the orange tab on the audio sink block. You should see a connection appear between the two blocks.

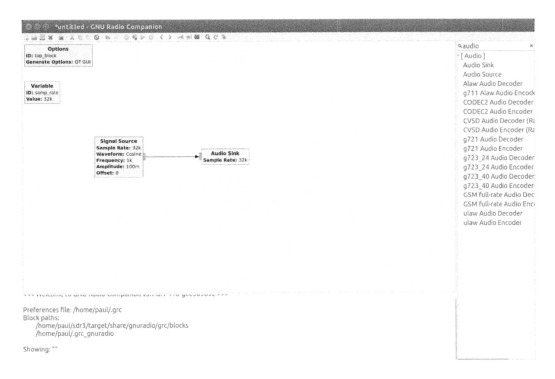

7) Click the Execute icon on the toolbar. Gnuradio will ask you to save the project at this point. Name the project verify.grc and click the Save button..

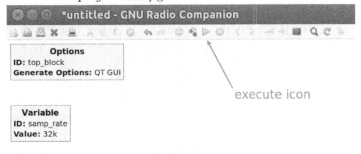

execute icon

8) After saving, gnuradio will take a second to generate and run your project. After a few seconds, you should hear a moderately-pitched tone. If so, then your gnuradio sanity check has passed! Let's move on to bigger and better things.

3 A Simple Radio Model

3.1 What is a Radio?

We all have an intuitive idea as to what a radio receiver is and what it does. At its simplest, a radio is a kind of magic box that picks up transmissions traveling through the air and makes sound from them. Breaking it down a little further, we know that those transmissions were broadcast from a transmitter some distance away, and may contain more than just audio. Maybe it was a radio or television broadcast tower 15 miles away. Maybe it was your laptop receiving wireless mouse transmissions from 10 centimeters away. Thinking back to science class, you might recall that those transmissions are electromagnetic in nature, and that they were described by some fairly complex equations.

I could keep diving down into the details and the theory, but in keeping with the layered onion philosophy of this book, let me propose a very simple model for a radio and then I'll proceed to flesh out a deeper understanding in successive chapters. To do that, I am going to start with possibly the simplest radio around, an AM radio.

3.2 Starting with an AM Radio

You have probably played with an AM radio at some point. It picks up transmissions sent out by a big radio tower somewhere in the vicinity and turns those transmissions into audio. We are going to spend most of this first volume working with the concepts underlying AM radio and use them to explain how software defined radios work. First, I am going to give you a model for the AM radio, and then we'll go a little deeper. Keep in mind, some of the terms will be a little unclear at first, but I'll progressively clarify them as we go through the chapter.

The AM radio system will include a transmitter as well as one or more receivers. Let's represent that with a simple block diagram.

Not so hard, right? Something is creating the signal, it's being sent through the air and something is receiving the signal. Let's break it down a little bit further. At the site of the transmitter, there is some audio that the radio operator desires to broadcast. This audio signal is then modulated onto a carrier wave and sent out as a radio signal. The receiver then tunes to the desired carrier signal and demodulates it to recover the original audio.

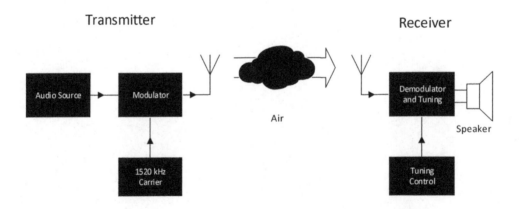

OK, there's a lot to unpack here, so I don't expect it to be perfectly clear yet. One thing might make sense though. Do you see the carrier frequency in the previous diagram? 1520 kHz is equivalent to 1520 on your AM radio's dial. So the audio you are listening to on your car radio illustrates an important concept. At various AM frequencies, you have sound somehow being transmitted. Radio engineers have found a way to take sound, and send it out to the world on different radio frequencies. And another thing: different folks can transmit at different frequencies simultaneously. In Seattle, for example, there's audio flying around at 570 kHz, 1150 kHz, 1300 kHz, and many other AM frequencies. How do they do it? And what does "modulation" even mean anyway? To see that, we'll need to take a closer look at the actual signals involved.

3.3 AM Radio Signals

A lot can be said about signals, but for now, just think of them as some physical property that varies over time. Maybe it's the air pressure that makes up the sound in the air. Maybe it's the voltage on a wire that goes to your radio's speaker. Maybe it's the electromagnetic intensity received by a radio antenna. In any case, they're all going to look generally similar when we sketch them on paper. They will have some kind of vertical axis representing the signal's value and a horizontal axis representing time.

For example, when it comes to sound, here is what the signal for a simple tone looks like. It just keeps oscillating back and forth for as long as the tone persists. You might recognize that from your trigonometry classes as a sine wave or sinusoid. For now, let's assume that the AM radio station is transmitting this simple sound.

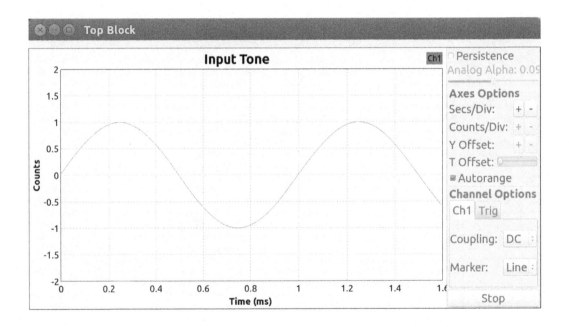

Now let's look at the carrier. It's actually very similar to our simple tone above, it just moves back and forth a lot faster.

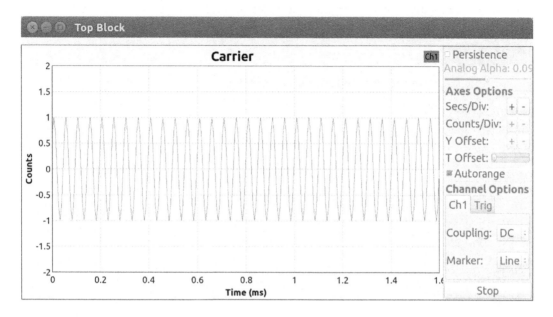

Now let's go back to the simple tone and modulate it onto the faster carrier. It is probably a good time to mention that AM actually stands for Amplitude Modulation. This simply means that we modulate (or change) the amplitude or strength of the carrier based on the audio signal. Specifically, it means we proportionally reduce the size of the carrier when the audio signal is low and increase it when the audio signal is high. In fact, you can see the original signal imprinted on the shape of the carrier. Take a look.

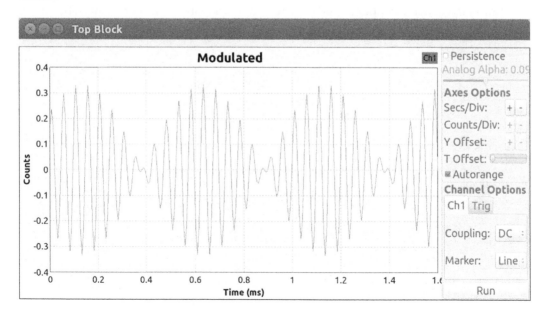

When a person starts talking, the resulting audio signal gets more complex as you can see here. But modulating this onto our carrier produces the same result, and again, you can see the outline of the original audio signal on the modulated carrier.

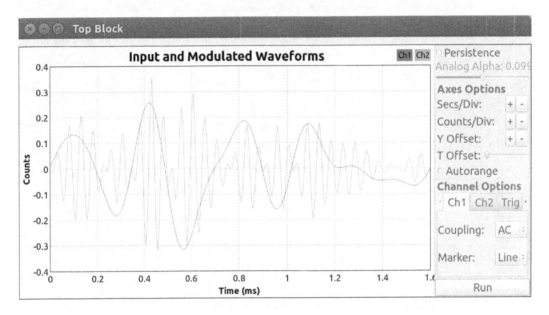

So now you have the basic idea behind modulation. You can see what it looks like in the images above, and later we will work on actually doing the modulation using gnuradio. I know you probably have a lot of unanswered questions, such as "why do we need a carrier?" or "how did we choose the carrier frequency?" or "what other kinds of modulation are there?" We'll get to all those questions and more.

But first, we need to understand how we can work with signals using our computer.

4 Computers and Signals

4.1 Computers and Signals

So we looked at a bunch of signals in the last chapter, and essentially they were squiggly lines moving around over time. This book is about software defined radios, essentially building radios with your computer, so the question you might have is "how do I get those squiggly lines into my computer?"

To answer that, we need to take a little excursion into the realm of digital signal processing. Not a long trip, because we need to stick to our goal of learning by doing. Just be aware that there is a whole field of study dedicated to digital signal processing, and we will only be scratching the surface here. But that's OK, because you can start your active learning with just a few concepts under your belt.

4.2 Digital Sampling

Let's start with how to get a real world signal into a computer. This signal could take many forms, but for now let's assume it's simply the voltage on a wire. At any given time, the voltage could be 1.3 V, or it could be -0.042 V, or it could be 110 V. What we need to do is measure that voltage and translate it into a digital value that a computer can understand. But remember, the signal will likely change over time, so we need to keep measuring and generating new values for as long as we want to look at the signal.

The hardware that does this is called an analog-to-digital converter, or ADC for short. It's called this because the real world signals are considered "analog" and the numerical measurements are considered "digital." Let's look at an example.

Possibly the simplest signal you will ever encounter is a square wave. It alternates periodically between two different values. In the case of the example below, it switches between 0 V to 3 V and back to 0 V, and it does so every second.

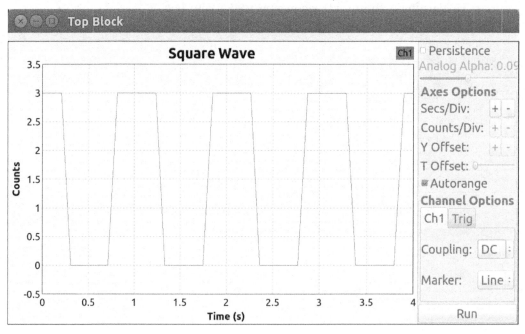

When we feed this signal to an analog-to-digital converter, it will produce a stream of 0s and 3s, typically displayed like this.

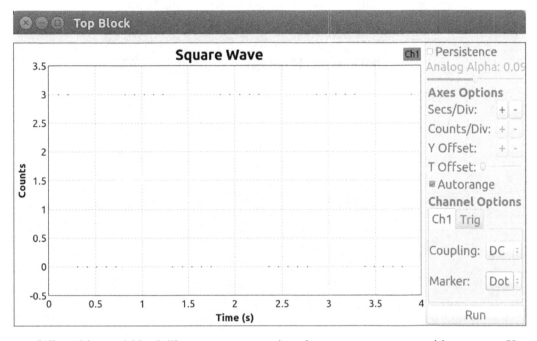

What this would look like to a computer is a data stream or array with contents [3, 3, 0, 0, 0, 0, 0, 3, 3, 3, 3, 3, 0...].

Now take a look at a sinusoid and its sampled version (we mentioned sinusoids back in the AM Radio Signals chapter). Sinusoids are an important waveform type with unique properties useful for our work - get used to seeing them.

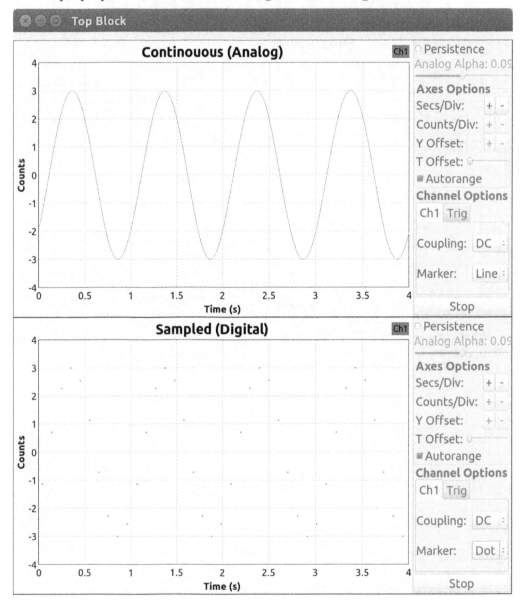

Just so you don't think ADCs are limited to artificial signals, take a look at a more arbitrary waveform, similar to the audio snippet we saw in the last chapter.

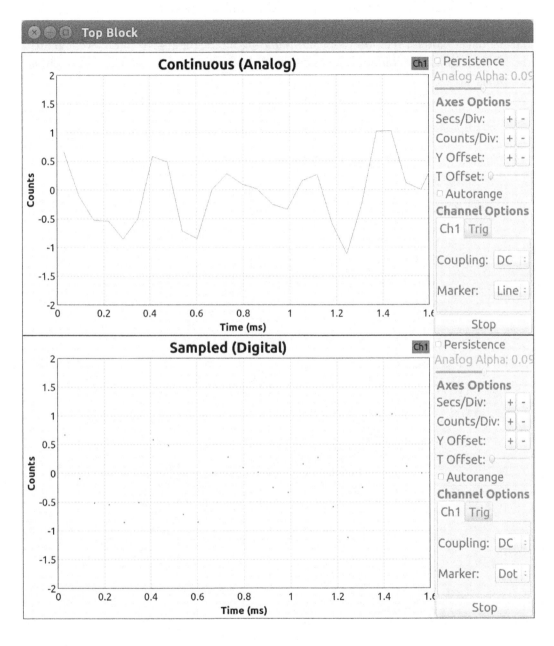

In this case, the number stream would be roughly [0.7, -0.1, -0.5, -0.55, -0.85, -0.5, 0.55, 0.5, -0.7...].

We can also flip this around and convert digital (computer) values to analog (real world) values. The thing that does this is unsurprisingly called a digital-to-analog converter or DAC.

In summary, an ADC measures stuff in the real world and translates it into something your computer can understand. The DAC does the opposite, allowing you to take the values your computer comes up with and send it out as a real world signal. As you probably expect, there are a lot of details missing in this simple description. I haven't talked about precision, noise, anti-aliasing filters, binary storage... we'll get to those things in a later volume. The one thing I do want to cover a bit more, though, is sampling rate.

4.3 Sampling Rate

Remember the plots of the sampled waveforms a few pages ago?

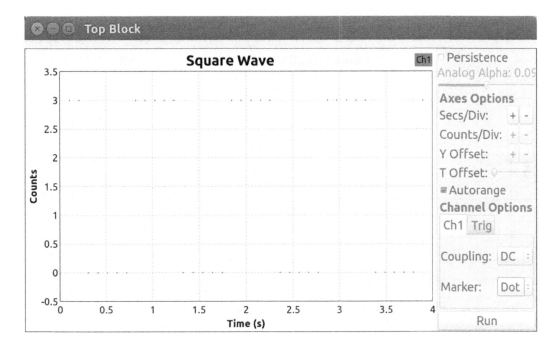

Did you notice how far apart the samples, or dots, were? The interval between them is called the sampling period. Conversely, the number of sample periods per second is called the sample rate (or sometimes sampling rate). You can think of the sample rate as how fast you measure the signal.

The most important question is then "how fast should I sample my signal?" Take a look at the following sinusoidal waveform that's sampled at a relatively high sample rate.

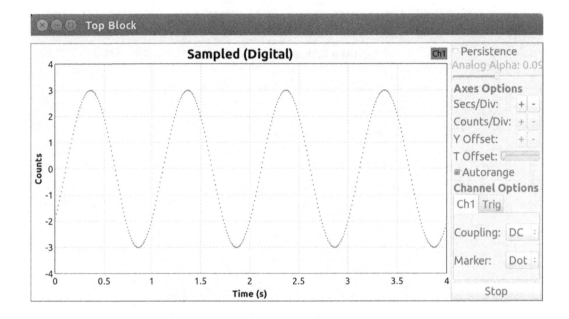

See how the original continuous signal is visible even to your naked eye? When your computer receives this data stream, it will have a faithful reproduction of the original signal.

Now let's see what happens when we sample relatively slowly. Again, here's the continuous waveform with the sampled one.

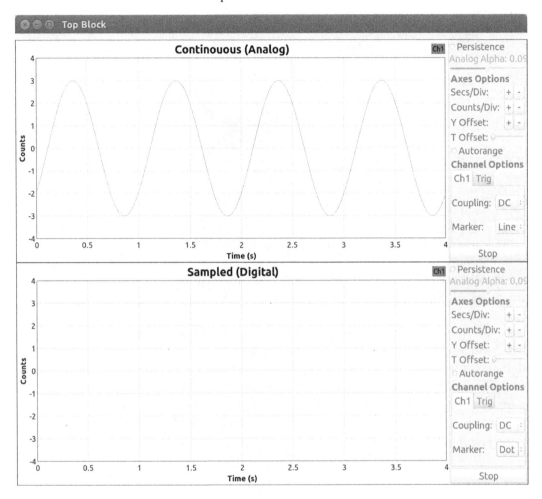

Can't really see the original signal anymore, can you? It's because we're sampling so slowly that we missed things. By "slowly," I don't mean that the sample rate is slow in an absolute sense, but that it's slow relative to the signal we're trying to sample. The sample period is so long that the quickly changing signal is able to move around too much between successive samples. In between each pair of samples, we essentially lose track of the signal.

Here's one way to think of it. Imagine that you are standing in a room with a very active cat. It likes to pace around in a big circle but somewhat randomly changes direction.

Clearly we went to engineering school, not art school.

For the sake of illustration, let's say the cat can stroll through a complete circuit of the room in 10 seconds at its very fastest.

Now someone turns off the lights. Then that same someone starts flipping the light switch up and down. Every second, they flip it on for an instant and then back to darkness. The result of all this is that every second you get a brief glimpse of where the cat is located. Since the light is flickering fairly quickly relative to the slow cat, you have a pretty good idea of the cat's motion. Even when it changes direction or speed, you can be sure you know where the cat is and where it has just been. It just doesn't move fast enough to escape being observed in the flickering light.

Next imagine that the time between light flickers increases significantly. Now the light switch only toggles to give you a view of the room every 60 seconds. Under these conditions do you think you have any idea how the cat is moving? In between those rare flashes of light, the cat could be doing anything: switching direction, speeding up, slowing down. When you "sample" the room so slowly, relative to the speed of the cat, you really have very little idea what the cat is doing. Can you see a similar thing happening with our undersampled sinusoid?

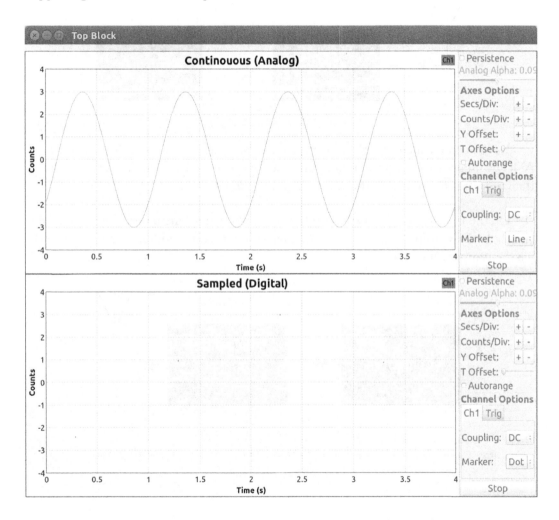

Here's the most important takeaway from this chapter: Sample fast enough or you'll get bad data.

But that just raises another question. How fast is fast enough? Answer: you need to sample significantly faster than the signal you're trying to measure.

Assuming our signal is not a simple sinusoid, how do we know how fast it is? Another great question. Don't worry, I will address this in the coming chapters.

4.4 SDRs from 50,000 feet

Now that you are armed with a basic understanding of analog-to-digital and digital-to-analog conversion, it's time to look at how an SDR works. Here's the simplest model that explains SDRs when they are used to receive radio signals.

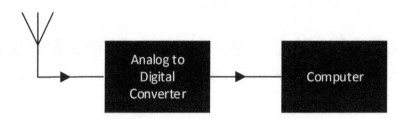

Not so complicated, right? Your receive antenna picks up some radio signals, the ADC translates them into something the computer can understand, and the computer then processes the signals to do whatever you would like.

On the transmit side, it starts with the computer generating a digital version of the signal, the DAC translating the signal to analog, and then broadcasting it to the world via the transmit antenna.

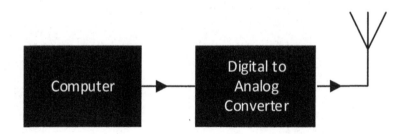

These models leave out a lot of details, but they are a good starting point.

The key here is that we have a computer at the root of both block diagrams. A computer that we can reprogram to do almost anything we can imagine. We can:

- extract audio from radio signals and play them back through our speakers,

- capture raw radio data to a file so we can analyze it later,

- change the SDR's programming so we can transmit or receive completely different types of signals whenever we need to,

- find mysterious signals and reverse engineer them,

- and much, much more.

In the next chapter, we'll start digging into gnuradio, the powerful software we'll use for capturing and processing radio signals using SDRs. It is gnuradio that will power the computer block in the previous block diagram.

OK. Enough theory. It's time to start learning by doing!

5   gnuradio and Flowgraphs

5.1 Flowing Streams of Numbers

In this chapter, are finally going to start playing with gnuradio. As I mentioned earlier, gnuradio is the application we are going to use to create new radios almost wholly in software. We'll click and drag various blocks onto our project, connect them, and configure them to build radios. Then we'll generate (an operation similar to compiling source code) and finally execute our code.

First, we start up gnuradio's graphical interface by typing the following into a terminal window:

gnuradio-companion

A split second later we see something like this:

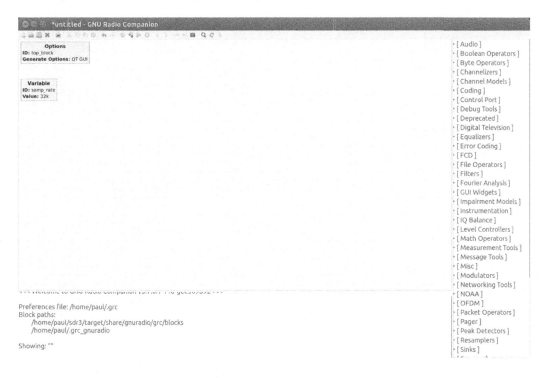

If you're familiar with LabView or other visual programming tools, gnuradio will likely strike you as very intuitive. If such programming tools are new to you, then let's take a few moments to go through how they work. Rather than writing our software in languages such as C++ or Python, with a host of text files containing such things as "if" statements and "while" loops, gnuradio-companion lets us build our program graphically. In fact, gnuradio-companion programs are referred to as "flowgraphs." As we work with gnuradio-companion more, the name will increasingly make sense.

When we build our flowgraphs, we will click and drag blocks from the right side of the interface to the main workspace area.

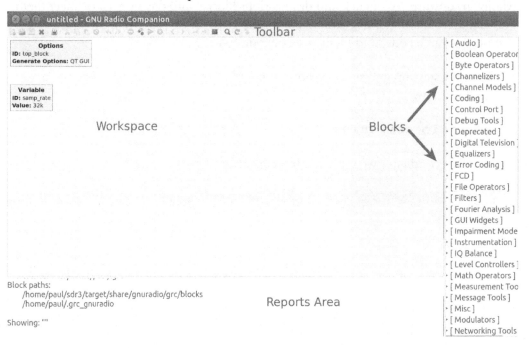

We will then connect the blocks in the flowgraph area in whatever way we choose. We will also configure each of the blocks depending on the particular needs of our flowgraph.

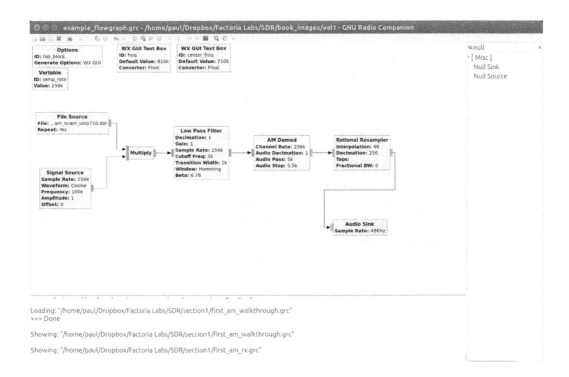

We will use blocks known as "sources" to introduce data to our flowgraph. The data will then be processed in some way before being sent to one or more "sink" blocks. The key concept is that data enters our flowgraph through sources and leaves through sinks.

5.2 Sources

Let's talk about the various ways we can get data into our flowgraph. Remember how the ADC in our SDR hardware produced a steady stream of numbers, based on the kind of radio signals being measured? We can generate this data stream using a source block that connects to our SDR hardware, which we will discuss in Volume 2.

But even if we don't have access to SDR hardware, there are still other sources we can use to input data into our flowgraph. For example, we can use a File Source block, which gets its data from a selected file on our computer. The File Source block is what will allow us to work through the projects in this book without requiring any SDR hardware. The authors of this book have generated a number of data files that simulate the behavior of real SDR hardware, so you can learn the basics much more easily.

There are also audio sources, random data sources, and a host of synthetic sources that let us generate pure sinusoids, among other things. Again, a source is anything that inserts data into our flowgraph.

5.3 Sinks

The counterparts to the source blocks are sink blocks. Data flows into our flowgraph via sources and exits our flowgraph via sinks. The sink blocks are really the outputs of the software defined radio we are building.

Some sink blocks are used to send data to our SDR hardware for transmission. We can also send data to an Audio Sink block so we can hear what it sounds like. Lastly, we can capture the data our flowgraph produces using a File Sink block.

Although many of the sink blocks have source block counterparts, one special type of sink block does not. These are the instrumentation blocks. The instrumentation blocks allow you to visualize your flowgraph's data in real-time, using a number of different techniques. Let's start with a little project right now where we use nothing more than a source and a sink.

First we create a new flowgraph and then double click on the Options block in the top left corner of the display. This block is automatically included in all gnuradio flowgraphs and contains some configuration settings. For this short project we'll be using a type of graphical user interface (GUI) called the WX GUI. If you look at the line labeled Generate Options you'll see that it is currently set to QT GUI. Clicking on the QT GUI text reveals a pull-down menu, from which we select **WX GUI**. In Volume 2 we'll see the difference between the QT and WX GUIs, but we won't dwell on this right now.

Properties: Options	
General Advanced Documentation	
ID	top_block
Title	
Author	
Description	
Window Size	1920, 1080
Generate Op	WX GUI
Run	Autostart
Max Number	0
Realtime Sch	Off

Cancel   OK

Next, we click the magnifying glass icon on the toolbar to bring up the search box on the screen's right-hand side. Although its possible to hunt through the list of blocks, we recommend using the search function when you know what you're looking for.

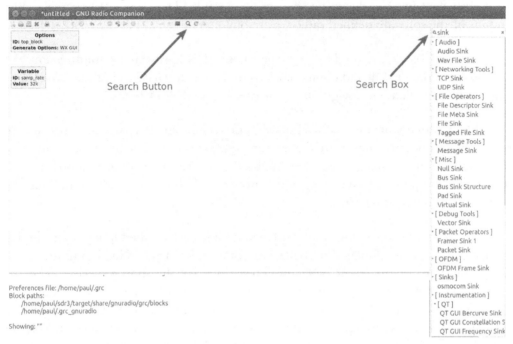

We then type **source** into the search box. The first block we see is a **Constant Source** block. After double clicking it appears in our workspace. This block will produce a constant stream of 0s.

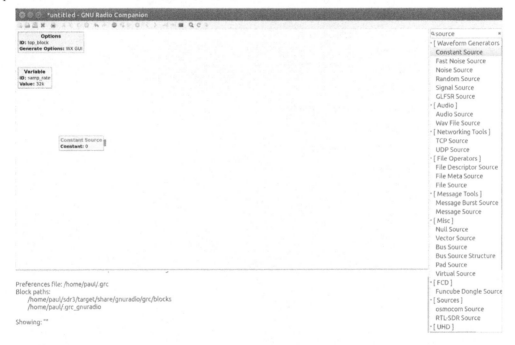

Next we replace the search box text with **sink** and then scroll down to something called WX GUI Scope Sink. After double clicking, it too appears in the workspace. This block will allow us to view a digital data stream in our flowgraph.

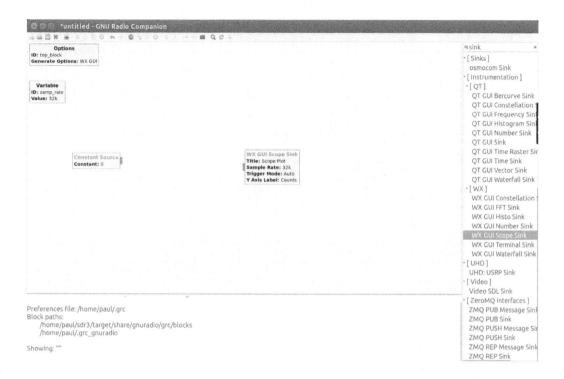

Note that you can move these blocks around however you want by clicking and dragging them. It is a good habit to organize your flowgraph neatly so it is as easy as possible to see what's going on.

We now click on the blue tab on the source block and then immediately click on the blue tab on the sink block. Do you see the connection appear? This is how you will connect all the blocks in your design. Notice that the connection is not just a line, but shows the data flow direction via the arrow. As we would expect, the data is flowing from the source to the sink.

Now it's time to run our simple flowgraph. We click on the Generate icon as shown below. We'll be prompted to save our flowgraph first, and we provide it the filename of simple.grc. If no errors are reported in the Reports window, we then click the Execute icon.

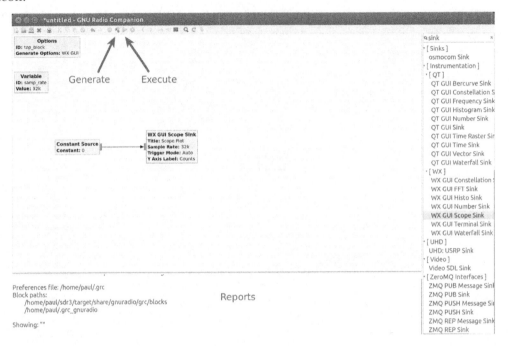

After a few seconds of chugging, gnuradio-companion brings up a scope window showing us a constant waveform with a value of 0.

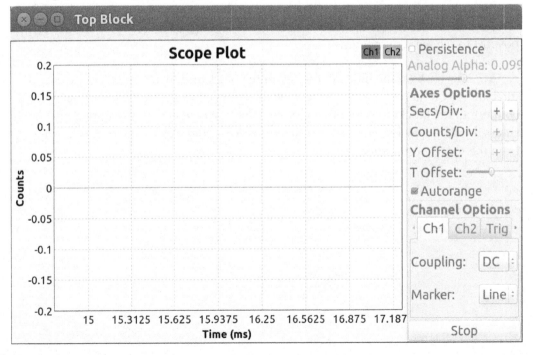

After a moment of gazing upon this beautiful waveform, we close the window by clicking in the upper left-hand corner. Closing the window will also terminate the simulation. If you tried to do something else while the simulation was running, you may have noticed it acting a bit sluggish. There's a fix for that, but we'll get to it a bit later.

As one last exercise, let's change the level of the Constant Source block. Double clicking on the this block brings up the following window:

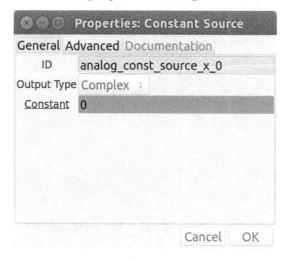

We change the value of Constant to **3.14** and then click OK. Notice how the block changed graphically to reflect the new value.

Executing the flowgraph verifies that the value has changed as expected.

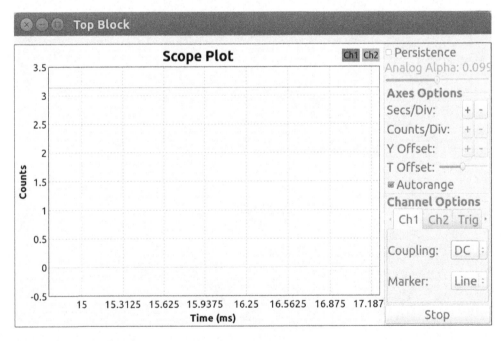

Sort of. The blue line (at 3.14) makes sense, but what about the green line at zero? What's on Channel 2, and why is it still zero? The short answer is that you've just had your very first complex number sighting in gnuradio. We'll see them again before long, but for now we won't get into the details.

5.4 In the Middle

So we know how to get data into our flowgraph, and we know how to get it out. What do we do in the middle? Well, the short answer to that is "math." Let's build on our last, very simple project (appropriately enough called simple.grc) so you can see how that works. After opening up the simple.grc file, use the File…Save As option to save it as simple_multiply.grc.

Next, going to add a Multiply Const block. This block will provide one of the simplest mathematical functions possible between our source and sink. First, however, we need to eliminate the connection that already exists between our source and sink. We can delete the connection in one of two ways: by right-clicking on it and selecting **Delete**, or by left-clicking to select and then hitting the delete button on our keyboard.

Now in the search box, we type **multiply**, and double click on the **Multiply Const** block.

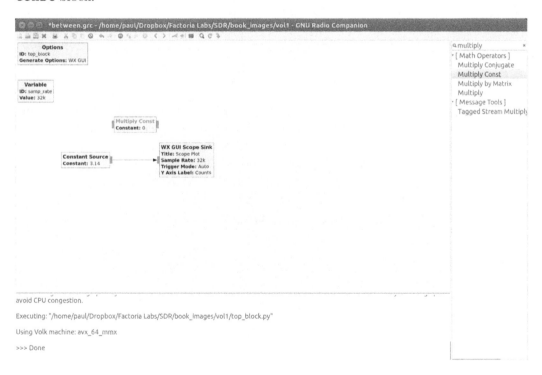

Next we connect the Constant Source block output to the input of the Multiply Const block, and the WX GUI Scope Sink block input to the Multiply Const block output.

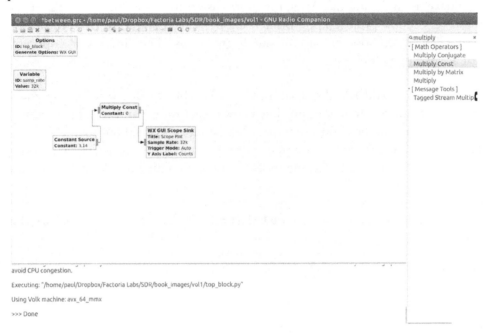

This particular Multiply Const block will multiply the incoming data by a constant value and then output it. The default constant is zero, but we'll change it to **2**. Remember that we double-click on a block to change its properties.

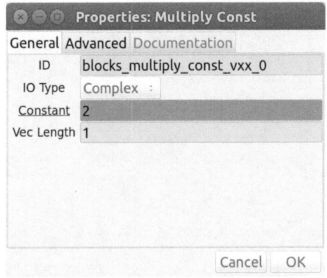

After clicking OK, we run the flowgraph again. Notice the new value is twice the previous one..

I know, multiplying by two is not exactly rocket science, but before long we'll be packing all sorts of interesting things between our sources and sinks

6 Your First Radio

6.1 Building the Radio

You probably have a lot of questions right now. Even though the stuff we've done so far has been fairly basic, you can probably see from the huge array of blocks on the right side of the gnuradio-companion interface that it gets more complicated quickly. Try to suppress those questions for a bit longer, because we're going to dive right into building our first radio.

There are two main things you probably need to learn right now: how to use the gnuradio software, and how radios work. I am going to focus most of my explanations in this chapter on gnuradio and how to use it. I will, however, mostly gloss over the radio theory right now. As such, some of the steps we take may not make sense yet, and it may feel like following a rote set of instructions, almost like building a model airplane. Rest assured, though, it is not my intention to give you simple cookbooks and send you on your way. We will keep coming back to this project through the rest of this volume to further dig into the details of how radios work.

6.2 A Simple AM Receiver

First, we create a new flowgraph by clicking File->New or by typing Control-N. This will bring up a simple starting flowgraph with two blocks in the workspace. The first is the Options block, where all of the top-level parameters of our project are stored. The other is a Variable block, representing the variable called samp_rate, which you might guess has something to do with the sample rate for the flowgraph. Every project will have both of these blocks.

Because we will be using the WX GUIs as we have before, we need to start by double-clicking the Options block and selecting **WX GUI** for the Generate Options selection.

Next we locate the block called WX GUI Text Box block using the search box feature. Then we drag it on to our workspace and drop it (you can also double-click it). This block will allow us to define a global variable that we can use throughout our flowgraph and also change in real time while the radio is running.

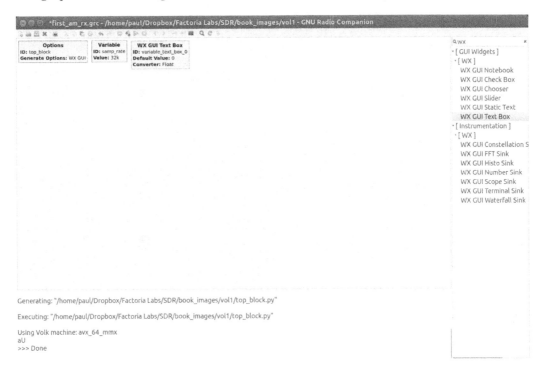

Double-clicking the new WX GUI Text Box block brings up its properties window, in which we change the ID to **freq** and set the Default Value to **880000**. Notice that we can use exponential notation **880e3** instead of typing all of those zeros. This is a common programming-language way of rendering 880×10^3, which if you do the math, is equal to 880,000. Remember this notation, as we will be using it a lot. Exponential notation is especially useful for large numbers, making them much easier to read. For instance, 600000000 is not as clear at first glance as 600e6.

When we click OK, note that the **WX GUI Text Box** block changes to reflect the values we have entered. This is useful because even as our radio flowgraph gets progressively more complicated, we can see a lot about how it works from a top-level view, all at a glance.

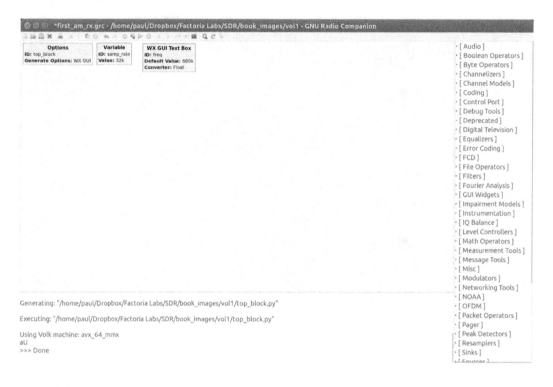

Next we will do the same thing again, adding a second WX GUI Text Box block with a different ID and Default Value. Instead of adding it like before, though, let's just copy and paste the first one. We simply click on the WX GUI Text Box block containing freq and press Control-C (or select Edit->Copy) followed by Control-V (or select Edit->Paste). Notice that the new block appears with freq_0 as its ID.

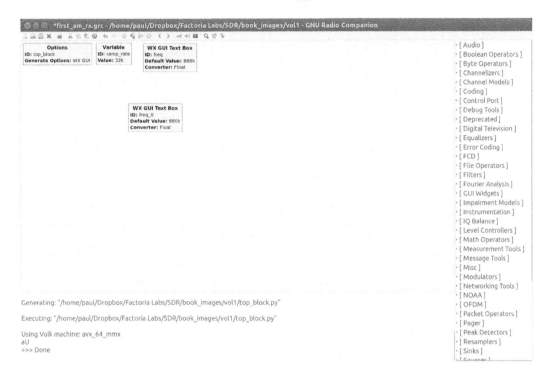

We will double-click this block and set its ID property to **center_freq** with a Default Value of **900000** or **900e3**. When you are done it should look like this.

We now grab a File Source block and add it to the design. This will allow us to use a file as the radio data input to our flowgraph, so we won't require any SDR hardware.

As before we double-click the newly placed block to setup its properties. In the File selection, we click on the three dots, then navigate to the location of the project files we downloaded earlier from www.fieldxp.com. Select the one named **am_ broadcast_02_c900k_s400k.iq**.

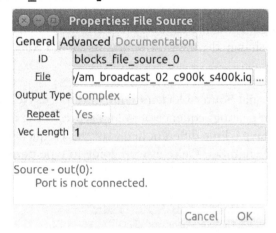

We then click OK, returning us to the flowgraph. Now that there is a source for our project, we need a sink and something in the middle. Note that if we were building an actual, working radio, we would use a different source block instead of the file source. This alternative source would interface with our SDR hardware and provide real-time radio data to our flowgraph. In case you are curious, the file contents here are just raw radio data that an enterprising SDR aficionado captured from the airwaves and stored in a file.

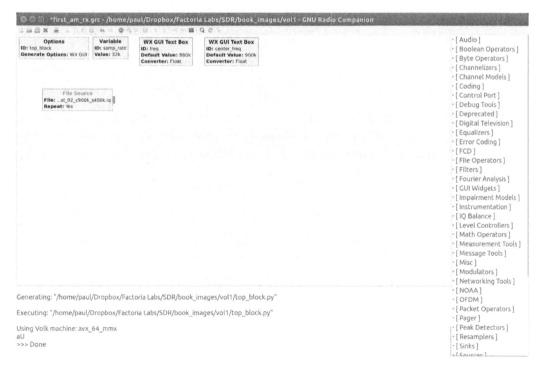

At this point, we should save our project by typing Control-S or by selecting File->Save. We can give it any legal Linux file name, but let's use first_am_rx.grc (rx is shorthand for receiver). Remember to save early and often.

NOTE: At this point, I are going to assume you know how to search for blocks and add them to your design. I will also assume you know how to bring up the properties listing for a block and change the relevant values. As such, I will stop spelling out each step of these processes.

We next place a Signal Source block into our workspace. We will use this to generate an infinitely repeating sequence of values representing a sinusoid. Then we are going to do something a little different here. For the Frequency property, we're not going to assign a simple number. We're instead going to type **center_freq-freq** into it.

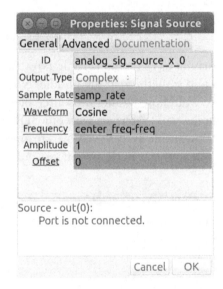

Can you see what we did here? Instead of entering a fixed number, we can enter variables. And not just variables, but mathematical expressions. We'll get into this more later, but you can actually enter almost any legal Python expression as a property and it will work. This will turn out to be very useful.

Also note that when we click **OK** and go back to our workspace view, we don't see the math expression we typed, but simply the number that results from it:

900e3 - 880e3 = 20e3

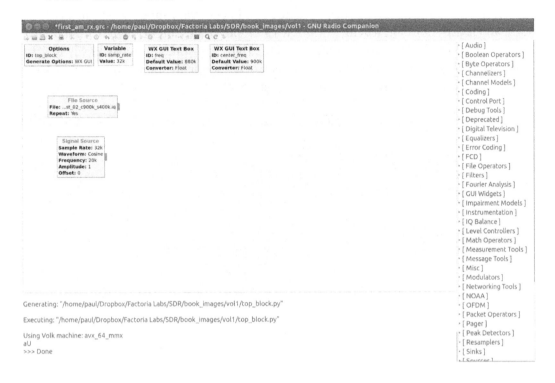

Next we place a Multiply block and connect its inputs to our two sources. Remember how to connect blocks? We simply click on the tab of the first block (in this case, one of the sources) and then click on an input tab of the second block (in this case the Multiply block). We can also do the clicking in reverse order and it still works. But which tabs are the Multiply inputs? Go ahead and hover your mouse over one of the tabs, and it will show you whether or not it's an input or an output. In general the inputs will be on the left and the outputs on the right, but sometimes blocks will be rotated and this will no longer be true. Don't worry too much about what this block is doing, we'll look at it more closely in a few chapters.

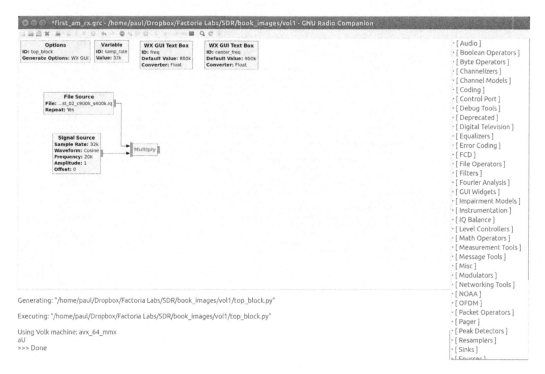

As we add and connect more blocks, it is good to take a moment here and there to tweak the positioning to neaten things. Just click and drag the blocks where you want them to go. As you do so, notice that the connections you've made are sticky and will follow your blocks wherever you drag them. This neatening process is not required, but can make your flowgraph much easier to read.

Next, we add a Low Pass Filter block, setting the Cutoff Freq to **5e3** and the Transition Width to **1e3**. You can leave all the rest of the properties alone. Filters are an extremely important concept that we'll spend time working with later in this volume.

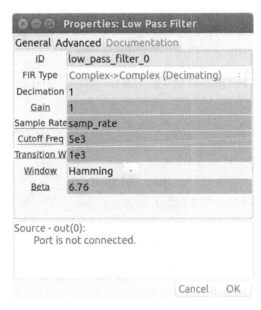

Its properties set, we connect the Low Pass Filter block input to the Multiply block output.

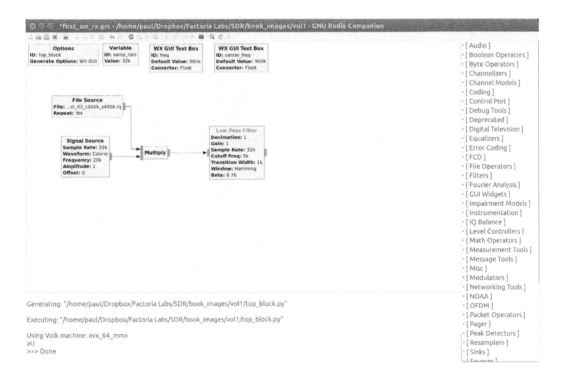

Another interesting point: do you see how the Low Pass Filter block's title is red, while the other blocks are black? (If you're colorblind, I guess the answer is no, but trust me, it's red). This happens because gnuradio-companion actively checks your flowgraph for errors as you build it, and any blocks with illegal conditions have their title displayed in red text. The illegal condition in this case is that the Low Pass Filter block's output is not hooked up to anything. Other common illegal conditions include missing or invalid properties, or multiple outputs connected together.

Now we place an AM Demod block, and set its Channel Rate to **samp_rate** and its Audio Decimation to **1**.

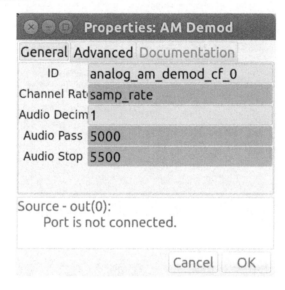

We now connect the AM Demod block input to the Low Pass Filter block output.

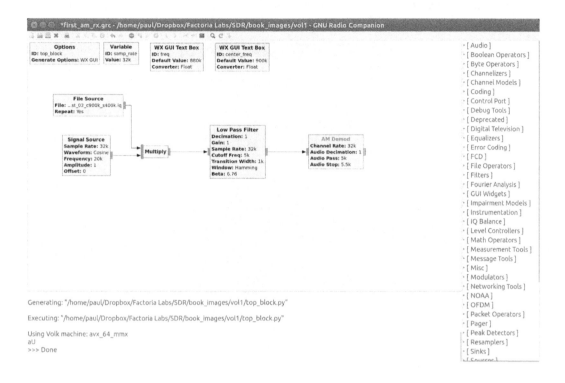

Notice how the input tab of the AM Demod block is blue, while its output tab is orange? This is significant, because the colors represent the type of data that flows into or out of the block. Orange tabs mean floating point numbers are flowing, while blue tabs mean complex numbers. Now is definitely not the right time to get into complex numbers (we are still on the outer layer of the SDR onion) so for now, just be aware that connections can only be made between tabs of the same color.

Next we place a Rational Resampler block and set the Interpolation to **32** and the Decimation to **400**. Then connect the Rational Resampler input to the AM Demod output.

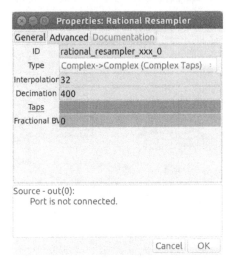

Again, it's probably not completely obvious what this block is doing, but we will take time later to dig deeper into the topics of resampling, decimation, and interpolation. Remember, at this point we're trying to learn the software and save the radio theory for later.

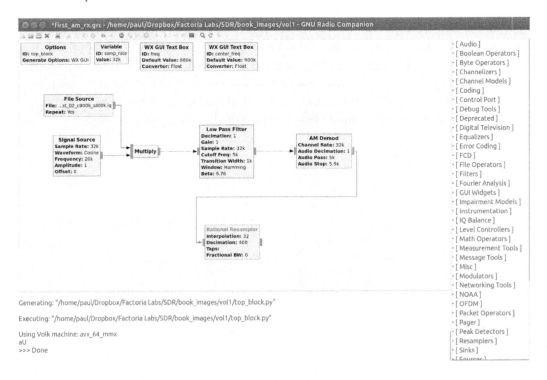

See that red arrow (sorry colorblind folks)? Something is wrong, and I gave you a hint a couple paragraphs ago. We can't connect tabs of different colors! Thinking in terms of programming languages, it would be like passing a string parameter to a function that's expecting an integer. So we open up the Rational Resampler block's properties again and note the Type property is Complex->Complex (Complex Taps). Instead we select **Float->Float (Real Taps)**. After hitting OK, we see that the colors of both tabs have changed to orange and the error is gone.

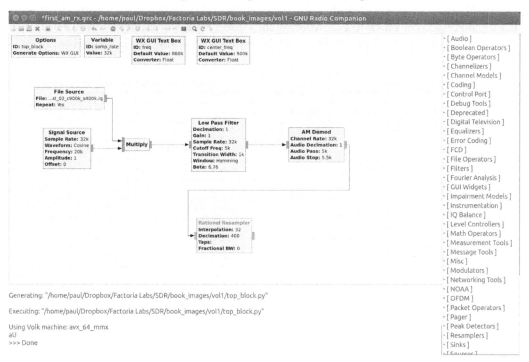

So we are almost done. Take a moment to think about what might be missing. We have a source (two of them actually), and we have some blocks that process the data coming in from that source.

Well, what are we going to do with our processed data stream? We need to dump it into a sink! We select an Audio Sink block which will "play" the data through our computer's sound card so we can hear the sounds being broadcast. We will use a value of 32 kHz (**32e3**) for the Sample Rate property and then connect the output of the Rational Resampler block to the input of the Audio Sink block. We could select a different sample rate, but the 32 kHz rate is commonly supported by most computer audio hardware.

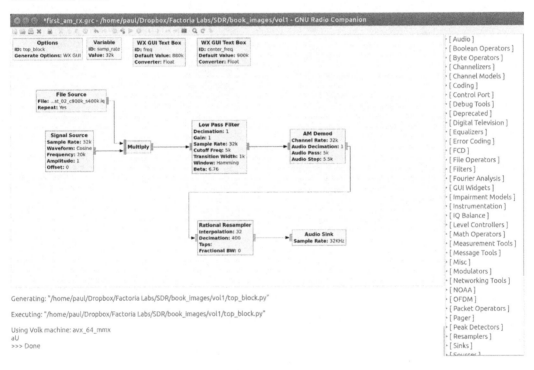

We want to do one last thing, which is change the sample rate for the flowgraph. Notice how most of the blocks have a Sample Rate of 32k displayed? This is because all of the blocks had a default Sample Rate property equal to samp_rate. If you remember, there was a Variable block present when we started our project. It had an ID of samp_rate and a value of 32k. When we double click on the samp_rate variable and change it **400e3**, do you see what happens?

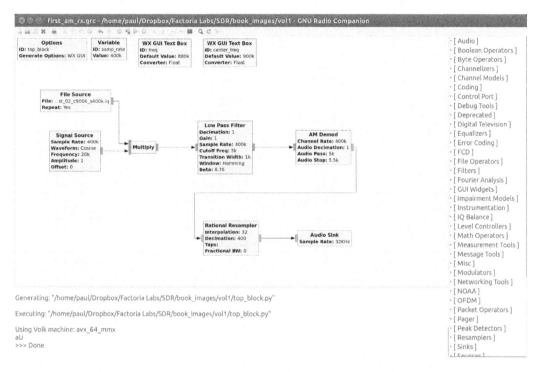

Changing a single Variable block caused changes to ripple throughout the design such that all the blocks now have a sample rate of 400k. All but one, anyway. More on that pesky Audio Sink block in a moment.

Let's run the flowgraph and see if it works! As before we just click on the toolbar icon that looks like a "play" button. If we hover over the button, it will say "Execute the flow graph." After we execute, some text will start scrolling by in the reports window pane.

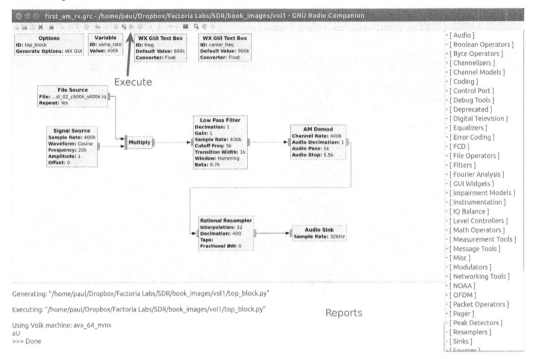

After a few seconds, we can hear a person's voice! There is a bit of static, as you may have heard on other AM radios, but there are clearly identifiable voices. The voices will loop after a few seconds, because there's not a lot of data in the file source and it's set to repeat. If you don't hear any audio, try adjusting the speakers on your computer.

The other thing you'll notice is that a window pops up with our two **WX GUI Text Box** blocks in it. As I mentioned early in this project, we can change these values while the radio is running. If we want to change any ordinary variables, like samp_rate, we would have to stop the flowgraph, change the value and run it again. However, if we change the value of freq to **750k** and hit enter, we immediately hear a different repeating chunk of audio. That's because we've now tuned to a different station. Go ahead play around with the freq variable and tune to several different stations. When you're done, you can kill the flowgraph by closing the variable window or by clicking the red x in the toolbar.

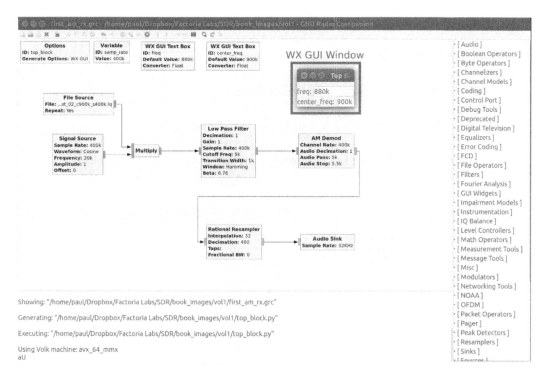

Think about it for a second. The person who created this input file didn't just record the audio for an AM station. That's pretty easy to do. They didn't just record the radio signal broadcast on a single AM channel. They recorded the signals on a whole bunch of channels so that we can tune to them at will.

This is your first glimpse of an extraordinary capability.

6.3 AM Receiver Overview

While not going into a lot of detail yet, let's peel back one layer of the onion on our radio design. Here's what's happening at a very high level. Our AM radio receiver flowgraph is:

1) injecting pre-recorded radio data into the flowgraph.

2) tuning to a specific AM radio channel.

3) filtering out other AM radio channels.

4) demodulating the signal of our desired channel.

5) doing some magical thing called resampling.

6) playing the resulting audio on our computer speakers.

You can see below the blocks responsible for each of these six tasks below:

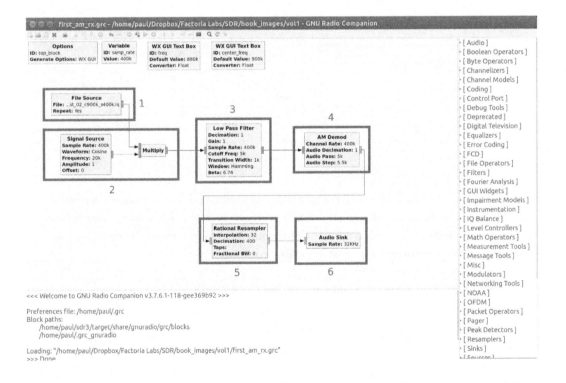

You probably have a few questions.

How does a filter work?

What does it even mean to resample?

How on earth does multiplying a radio signal by a sinusoid tune anything?

We are going to answer these questions throughout the rest of this volume, partly so you can see how the AM radio works, but even more so because the answers will illuminate some crucial radio concepts.

7   Frequency, Gain, and Filters

7.1 Frequency

When I say "frequency" a number of things probably jump to mind. Perhaps you think of a radio station frequency. Maybe you think of the specifications of your wireless router or mobile phone, probably in some numbers of megahertz or gigahertz. But there's a very useful occurrence of frequency that most of us have a pretty good intuitive handle on: sound.

You may remember that the human ear can detect sounds roughly between 20 Hz and 20 kHz (depending on age and number of heavy metal concerts attended). Higher frequency air vibrations than that still exist, but we call them ultrasonic. Your dog can hear these frequencies but not you. Even within the audible range, there are low pitched (aka low frequency) sounds that are like a low rumble. There are also high frequency sounds like a piercing shriek. There's also a whole range in between.

So when I give you a physics definition of frequency - the number of oscillations of a periodic phenomenon per unit time - this definition can apply to sound waves as well as to radio waves. The unit for frequency, Hertz (Hz), has a definition of one cycle per second, but that definition doesn't specify the thing cycling. It's just as feasible to have a 10 kHz sound wave as it is to have a 10 kHz radio wave. I am taking the risk of belaboring this point because you probably have some valuable intuitions about frequency, and I want you to know that they largely apply to radio concepts as well.

So getting back to audio frequencies, we could graphically represent the range of sounds as something like this:

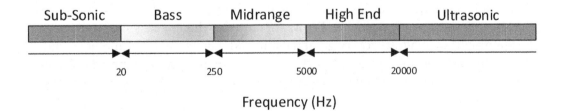

Frequency (Hz)

Interestingly enough, we can use gnuradio to work with audio frequencies in addition to radio frequencies. The software sees the audio data as nothing more than a bunch of numbers and has no way of knowing that the numbers did not come from a radio. So lets go back to gnuradio-companion and pretend for a moment that it's actually gnu*audio*-companion.

We'll start a new flowgraph and place a Signal Source block, setting the Output Type to **Float**, the Frequency to **500**, and the Amplitude to **0.1**. Then we drop down an Audio Sink block and connect the two blocks. Let's call this flowgraph single tone. grc. (Note that I am no longer providing screenshots of the entire gnuradio-companion interface but only the workspace portion of it. This makes it easier to see the details in our increasingly complex flowgraphs.)

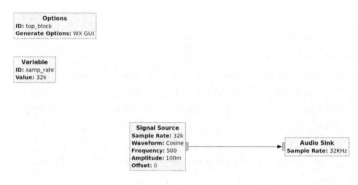

When we execute this flowgraph, you should be able to hear a low hum. If you cannot hear anything, try using headphones with the computer as the speaker may not be able to play the tone. This hum is the sound of a 500 Hz tone. Thinking of the above Audio Spectrum illustration, this should be unsurprising. It's a fairly low-pitched tone, but not the lowest imaginable. Before we move on, though, let's start probing the flowgraphs we built. It's a very good habit to add different types of monitors and displays throughout your flowgraph, so you can be sure that you're getting the signals you expect at each point. Thinking along those lines, we add a WX GUI Scope Sink block to the design.

NOTE: There are two different types of GUIs available: WX and QT. For this exercise, we'll be using the WX type of GUI. The Options block (usually in the upper left corner) will display the option setting. If the option setting displays "QT GUI" then we need to change this to **WX GUI**.

Going back to the WX GUI Scope Sink block, we set the Type to **Float** and connect it to the Signal Source output. Notice how we can attach the same output to multiple block inputs?

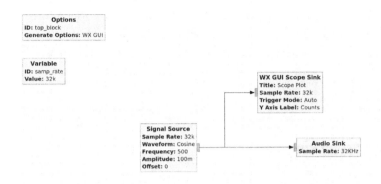

When we execute the flowgraph again, we see a Scope Plot window like this.

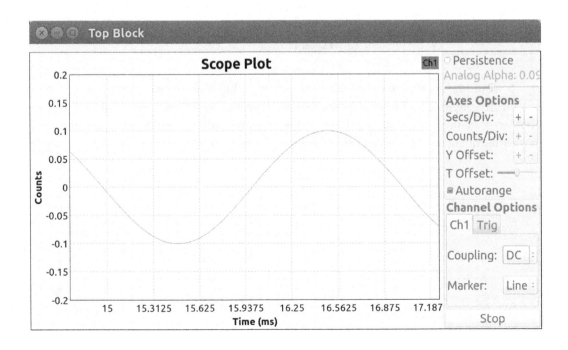

What we have here is roughly one period of a sinusoid, though if you're the suspicious sort, you may want more evidence of this assertion. If we zoom out a bit, we'll be able to see more than just the single oscillation. When we click on the '+' button next to Secs/Div: our view starts to zoom out. After two clicks we can clearly see several periods of the waveform thereby assuring ourselves that we do indeed have a sinusoid. You might be wondering why I chose a sinusoidal waveform to make this tone. The answer to that is long and mathematical, so I am going to ask you to take this one on faith for now - a pure, single-frequency tone is sinusoidal in nature.

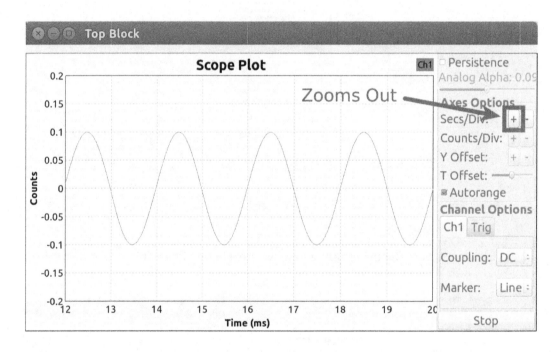

It would be good now to listen to a number of different tones to get an idea of the various frequencies and what they sound like. It would be easiest if we could adjust the frequency of the tone while the flowgraph is running rather than stopping, editing, and restarting the flowgraph with each change. Any ideas on how we could do that?

The WX GUI Text Box block we used previously provides just this functionality. We add one to the flowgraph, changing the ID to **freq** and the Default Value to **500**. Then in the Signal Source block, we change the Frequency to **freq**.

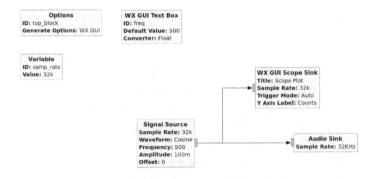

Now when we run the flowgraph, we have a single window containing the Scope Plot and the frequency control.

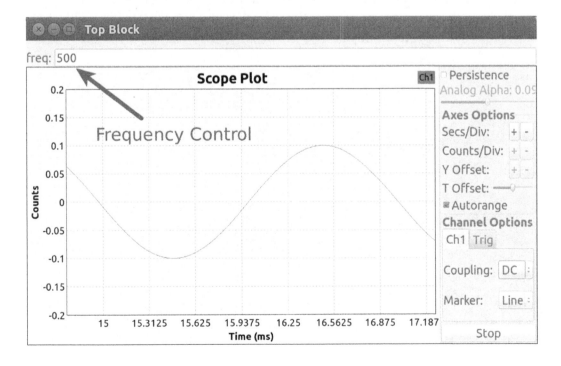

Now we can go ahead and change the numbers in the freq box (remember to hit enter after making a change) and then watch and listen. As the frequency values change, it might be necessary to zoom in or out to get a better look at the waveform. Also, your ear may hear down to 20 Hz, but that doesn't mean your sound card/speaker can output frequencies so low.

Does the behavior of your flowgraph make sense? Depending on how thoroughly you played around with the freq values, the answer is "sort of." When we enter numbers from about 100 to 16,000, the behavior does make sense. The tones increase in frequency and the sinusoid oscillates faster (or goes up and down more times in the same time period). Depending on your sound card's specs and the sensitivity of your ear, you may stop hearing things around 14,000 Hz or 14 kHz. If you keep going higher, however, you start to hear things again. And this time the sounds are progressively lower pitched the higher we raise the frequency. What's going on here?

One clue is to look at the sample rate of our blocks. Remember how I said that the sample rate just needs to be fast enough relative to the signals being sampled? Well, all of the blocks in this flowgraph are sampled at 32 kHz, and the funny business starts when our sinusoid is approximately half that value. We won't do the math yet, but now we have some evidence that sampling twice as fast as the signals in your flowgraph is a good idea. At least twice as fast.

At this point you might be thinking, "so what?" We've made some annoying sounds and looked at some squiggly lines. What does this have to do with radios? Well, as you'll see on the next page, quite a bit. Because we're about to talk about possibly the most important concept in signal theory. Welcome to the Frequency Domain.

7.2 The Frequency Domain

So let's get away from using synthetic tones and start thinking about more natural sounds. Most of the sounds we encounter in normal life don't seem at all like the pure tones we just generated, they sound fundamentally different. And yet there's still this idea of frequencies that are low - like a thumping subwoofer. There are also frequencies that are high, some so high we can't even hear them. If that rumbly bass sound coming out of the subwoofer isn't a pure tone like we've been making, then it must be something else and yet still have some kind of frequency characteristics (it sure sounds low, right?). Also, recall that some sounds are low and high at the same time. When you're listening to music, you can often hear the lower-pitched bass portions at the same time as mid-range singing and higher pitched instruments.

Here's the big idea: "natural" sounds are made up of a whole bunch of different frequencies that are simultaneously in play. And fortunately for us, we have mathematical tools that can break down a given sound into its constituent frequencies, showing what it looks like in the frequency domain. And not just sounds, any kind of signals. This mathematical tool is called the Fourier Transform.

We're too high up on the outer skin of the onion to talk about the mathematics behind the brainchild of Joseph Fourier, but here's a general explanation. We take a numerical representation of a signal, which you'll recall is just some value changing over time, and apply a mathematical function to the signal to create a breakdown of all the frequencies that make up that signal. Understanding how this works is crucial to understanding SDRs, so we're going to work through several examples until things are clear.

Strictly speaking, Fourier transforms are for continuous signals - remember the signals with smooth curves? There is a counterpart called a Discrete Fourier transform (DFT) that is used to do the same thing for sampled data, or the kind of digital data used in the SDR world. Because there are a number of ways to do the math for the DFT, we further specify the most common algorithm - the Fast Fourier transform (FFT). So this is the main takeaway: we take a continuous signal, sample it, run it through an FFT, and we find out what frequencies are present in that signal.

Before I ramble on any further here, let's take a look at the following. Starting with our last project (single_tone.grc), let's first rename it as single_tone_fft.grc, add a WX GUI FFT Sink block to our flowgraph and attach it to the signal source. The connection's red arrow is another warning that our input and output types do not match. After changing the WX GUI FFT Sink block Type parameter to **Float**, things should look good.

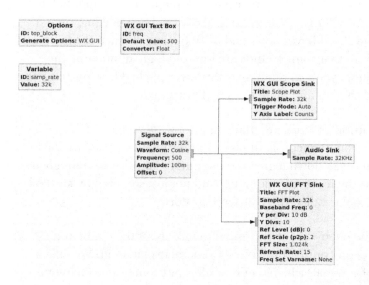

Executing the flowgraph brings up an FFT plot in addition to our existing scope plot and text box. Notice how the FFT shows a spike at 500 Hz and is non-existent for all other frequencies. This is experimental validation of my earlier assertion that sinusoids produce pure tones, or pure frequencies. Take a second to think about what you're looking at. One window is showing your signal in the time domain, the other in the frequency domain.

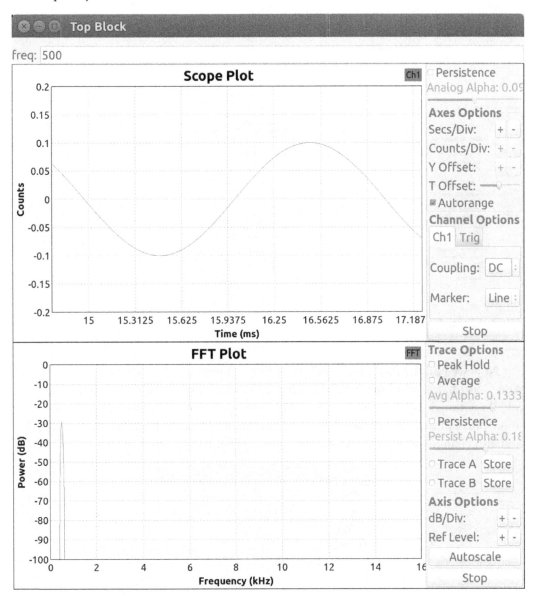

Now let's look at a sound with a bit more complexity. We'll rename the single tone_ftt.grc flowgraph as cmajor.grc. Next we'll delete the Signal Source block, and in its place we'll use three signal sources of different frequencies and add them together. The easiest way to do this is add the first Signal Source, set its Type to **Float**, its Amplitude to **0.1** and then copy/paste it twice. We'll set up the first Signal Source block's frequency (in Hertz) to **261.6**, the second Signal Source's frequency to **329.6**, and the third Signal Source's frequency to **392**.

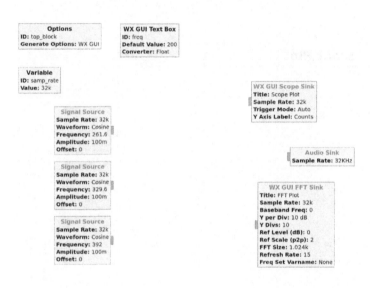

Now we find the Add block and add it to the flowgraph (be careful not to use the Add Const block by mistake). See how it only has two inputs? Since we are trying to add three things together, it would be nice if the block could accommodate us. Looking at the block properties, we can see a parameter called Num Inputs. We change it to **3** and also change the IO Type to **Float**, and behold! It's the block we want!

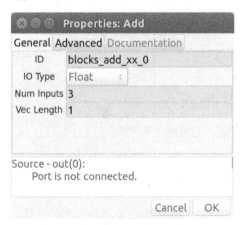

See how the Add block has changed such that it now has three inputs instead of the original two? We can now connect all three of our sources to the same Add block, which we do. We also connect the Add block output to the FFT, Scope and Audio Sinks.

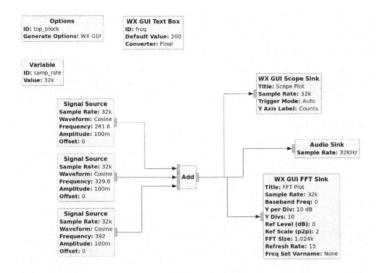

Let me make a brief aside here. Notice how the blocks that have both Float and Complex functionality (for the Type parameter) all seem to default to Complex when we place them into our flowgraphs? This is a clue to the fact that most of the time, we will be working with complex numbers. Slicing deep into the onion to get to the theory of complex numbers will only bring tears at this point (the onion analogy never gets old), so we'll keep to our gradual peeling. But be aware that complex numbers are in there, and that we'll learn more about them eventually.

Back to the flowgraph. When we execute it, you'll notice a more complex sound. This is not shocking since we're now putting three tones together. Use headphones if you cannot hear anything from the computer speaker. For your information, these are actually the tones that make up a C Major chord. When we zoom out on the scope plot, we can see that the signal is no longer a simple sinusoid, but seems to move around somewhat randomly. The FFT is now showing a peak, but it's wider and has some bulges on either side. The reason we don't see three distinct peaks is due to the limits of the FFT's resolution. The frequencies are just too close together.

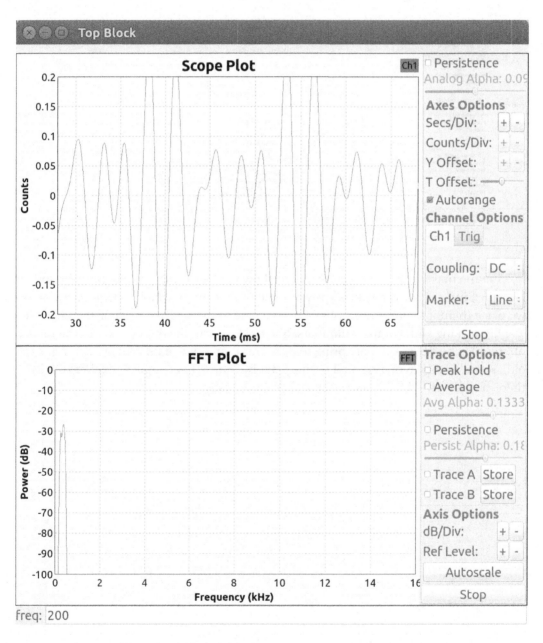

So let's fix that. After killing the flowgraph we change the frequencies of the second and third Signal Source blocks to 659.2 Hz and 1568 Hz respectively. We'll leave the first one at 261.6 Hz.

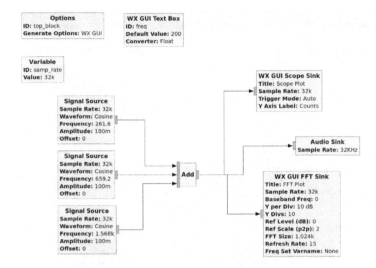

When we run this flowgraph, we can now see three distinct peaks. (Again, for the musically curious, we've just raised the second note by one octave and the third by two octaves.)

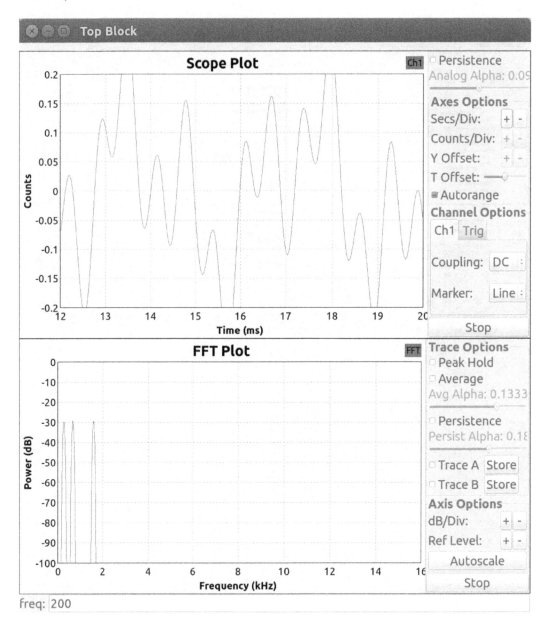

Remember the earlier assertion that natural sounds are made up from a bunch of different frequencies? Let's prove it. We'll save the cmajor.grc flowgraph as voice_fft. grc. Then we'll get rid of all the Signal Source blocks and the Add block by selecting all of them (left click and drag to select) and then pressing the delete key (or right clicking and selecting delete). In the place of these deleted sinks we'll add a Wav File Source block and connect it to each of the three sinks. Finally we open the properties for the Wav File Source block and navigate to the location of the project files we downloaded earlier from www.fieldxp.com. Select the file named HumanEvents_s32k. wav).

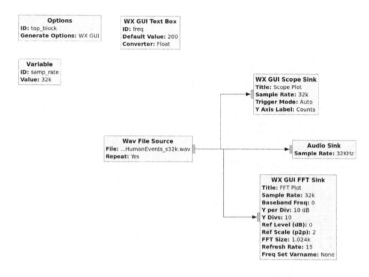

When we execute this new flowgraph, we hear a voice repeating over and over. We also see a fairly different view in the WX GUI Scope Sink and WX GUI FFT Sink blocks.

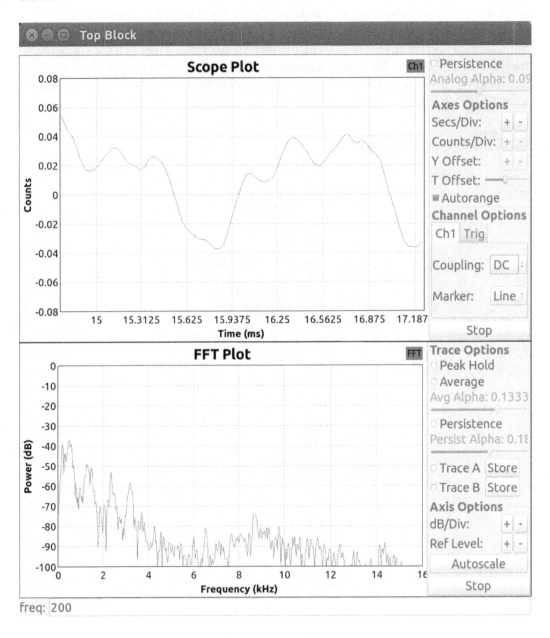

The Scope Plot shows a great deal of random movement, but the most important thing here is the FFT Plot. Look how spread out the frequencies are in a simple recording of a human voice. We can see that most of the signal is comprised of low to medium frequencies, but there are a few higher ones in there as well. The FFT Plot also moves around a fair amount, shifting while jumping up and down. If you want some extra credit, try recording some sounds of your own in WAV format and see what they look like. You will need to configure your audio recording software to use a 32 kHz sampling rate so your WAV file will work with this flowgraph.

It may not be obvious how exactly the FFT gets its job done, but hopefully it's becoming clearer what that job is and how important it is. From now on, when we think about signals, we will often be thinking about the frequencies present in the signals.

But we want to do more than just look at stuff, right? We also want to be able to change the signals we encounter. There are many ways to process signals, but we'll start by looking at two of the most foundational methods: gain and filtering.

7.3 Gain

Gain is a relatively simple concept that you may already understand better than you realize. This property really just represents how much bigger you make a signal. If you have a signal with an average size of 3 (this will work with any units), and you increase the size to 6, then your gain is 2. If you were to increase it to 9, your gain would be 3. Mathematically, we can express this as:

$$Gain = \frac{OutputSize}{InputSize}$$

Not so hard, right? Well, there is one snag: how do we define how big a signal is? If we had a square wave or sinusoid with a peak value of 2, then it might be pretty straightforward. But what about the signal in the last chapter that represented a human voice? How do we assign a size to that? It was squiggling all over the place when active, and then at times it was nearly silent altogether.

There are a number of answers to this question, based on how you want to define it, but we'd like to put off that particular issue to a later volume. For the moment, consider that applying a gain to a signal will increase that signal at every point in time by that same multiplier. Let's open a new gnuradio project to see this firsthand. We'll save this new project as gain.grc.

We start by opening the Options block and changing the Generate Options property to **WX GUI**.

Next, we add a Signal Source block and set the Output type to **Float**.

Then we add a Multiply Const(ant) block, setting the IO Type to **Float** and the Constant property to **gain**.

Next we insert two copies of a WX GUI Scope Sink block and also set both of their Type properties to **Float**.

Now that all of our blocks have been placed we can commence with hooking them together. We start by routing the Signal Source output to the Multiply Const block input. Then we attach one of the WX GUI Scope Sink blocks to the output of the Signal Source block and the other WX GUI Scope Sink block to the output of the Multiply Const block.

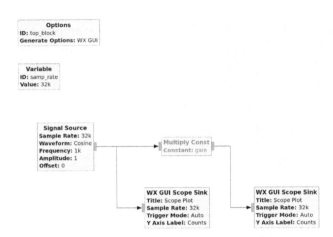

You can probably see what's going on here. The sinusoid generated by the Signal Source block is going into a block that's going to multiply it by whatever the value of "gain" happens to be. Then we're going to use a pair of Scope Sinks to look at the signal before and after the multiplication process. For this we need to provide a value for the gain variable. We'll use a new component called a WX GUI Slider block. After placing it, we set the ID to **gain**, the Default Value to **1**, and the Maximum value to **10**.

Finally, we change the Title property of the WX GUI Scope Sink block connected to the Signal Source block to **Input**, and the WX GUI Scope Sink block connected to the Multiply Const block as **Output**. It's generally a good idea to label your instrument blocks if you're going to have more than one of them. If all of your WX GUI Blocks are named "Scope Plot," you may find it difficult to determine the signal contained in each display.

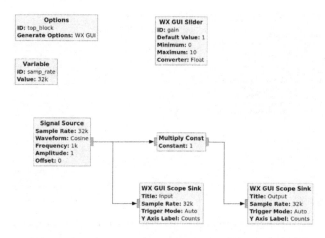

After executing the flowgraph, we see two identical sinusoids. First we deselect the **Autorange** checkbox on the Output scope. This allows us to smoothly see the effects of the gain slider. If we leave the Autorange feature enabled, the WX GUI will automatically change the y-axis scale to match the size of the signal in the display. Autoranging is often helpful, just not in this case.

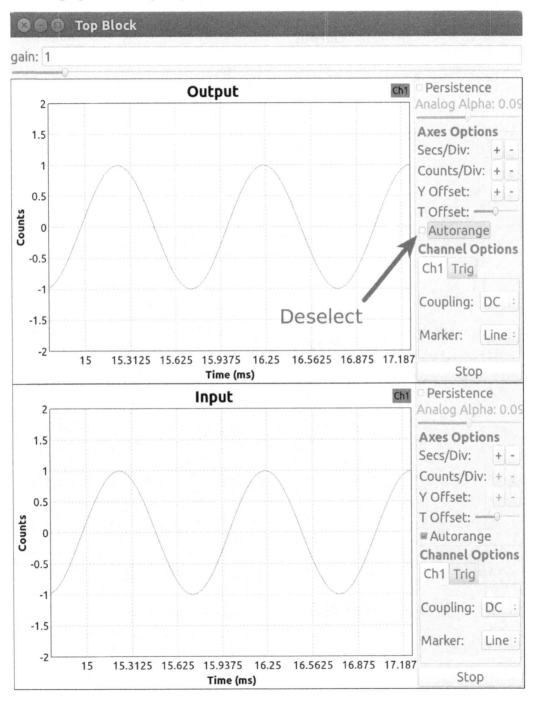

Next, we click on the gain slider near the top and drag it to the right. See how the signal increases in size?

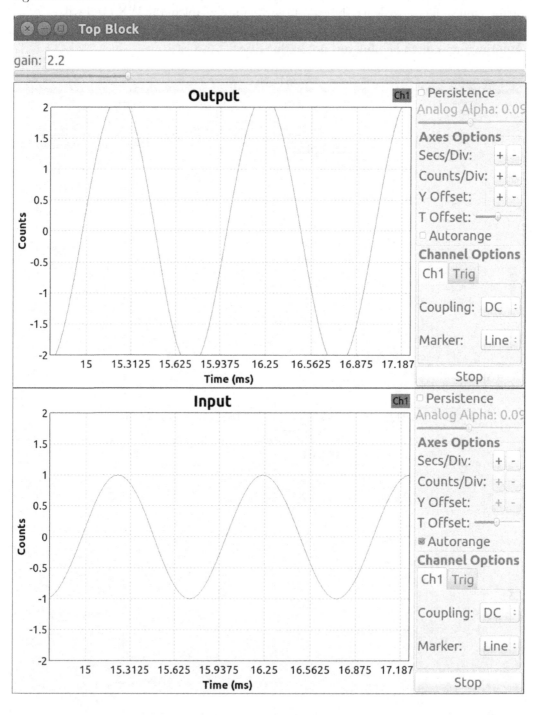

Can you also see how it decreases when the slider is moved to the left? I didn't mention that before, but while gain can be greater than one (making the output signal bigger), it can also be less than one (making the output signal smaller). When the gain is less than 1, this is typically called attenuation. For example, we can move the slider down to 0.1 to see the signal shrink by a factor of 10.

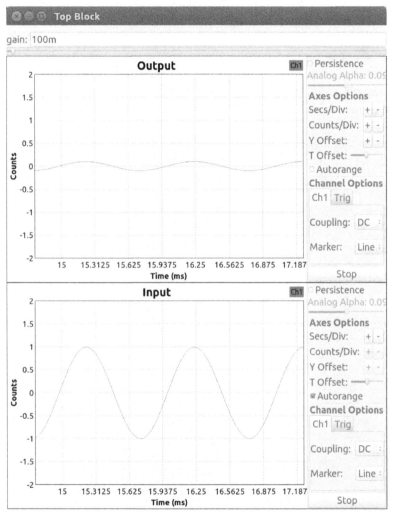

Note that when we make this change the number displayed by the WX GUI Slider is actually 100m. The 'm' here stands for the metric term "milli" and means thousandths. As we use the gnuradio GUI more, we'll see a number of other metric terms such as:

    u (micro or millionths)
    k (kilo or thousands)
    M (mega or millions)
    G (giga or billions)

Another note on terminology - blocks like our Multiply Const block, that provide only gain to a signal, are typically referred to as amplifiers. Be aware, though, that many other types of blocks may have some gain or attenuation associated with them.

There is one more thing I want to mention. Do you remember in our previous usage of a math-related block, where we generated the C Major chord, that we used a 3-input Add block? This time, our Multiply Cont block had only one input. Can you see why there was such a difference? In the first case, we were combining three different signals, or streams of numbers, and adding them together at each point in time. In the second case, we are only concerned with multiplying the signal, or data stream, by a single constant number. If you look at the library of blocks available, you'll see that there is also a Multiply block as well as an Add Const(ant) block.

If you're feeling adventurous, go ahead and substitute the WAV file from the previous chapter (or one you recorded yourself) for the signal source.

7.4 Decibels

Before moving on to filters, we need to throw a bit of a monkey wrench into the works. Ever heard of decibels, or dB for short? Gain is most often measured in decibels, and if decibels are new to you, this may complicate things just a bit. The main thing is to keep in mind that decibels are a logarithmic measurement, similar to the Richter Scale for measuring earthquakes. As you might recall, an 8.0 earthquake is not just 14% stronger than a 7.0 earthquake, it is ten times stronger. The same 10x multiplier applies to each change of 1 on the scale: 3.0 to 4.0, 5.7 to 6.7, etc.

Now, when gain is expressed in decibels, each incremental step up represents a much larger increase. We can illustrate this with a simple change to our previous gain project. Before we go any further, , though, lets talk about some sluggishness your computer may be experiencing when running these flowgraphs. It turns out that these sample rates we specify in our gnuradio flowgraphs are not exactly what they seem. For the 32 kHz sample rate in our current flowgraph, we would expect our blocks to complete an operation once every 31.25 microseconds. We compute this as:

$$\frac{1}{32kHz} = 31.25\mu s$$

A modern computer can complete your flowgraph's calculations much faster than that, so what does it do when it's done? Well, when we run a flowgraph that contains an interface to the physical world, such as an SDR-interfacing block or an Audio Sink block, the computer will sit around idling until it's time to process the next data sample. But when everything in our flowgraph is synthetic, something a little strange happens. Since there's no SDR hardware or sound card that's expecting data at any specific point in time, gnuradio doesn't really need to wait for anything. It's essentially just running a simulation, so why not run it as fast as possible? And so your computer does, greedily hogging resources you'd like to use for other things like moving your mouse around and clicking on things.

The solution to this is simply adding a Throttle block. This will force data at a certain point in the flowgraph to slow down to the rate specified by the Throttle block. To do this, you just need to add the Throttle block immediately after one of your source blocks. In our specific flowgraph, we delete the connections from the Signal Source block to the rest of the flowgraph. We then add a Throttle block, change its Type to **Float** and note that its sample rate is conveniently 32 kHz already. We then connect the Signal Source block output to the Throttle block input, and finally the Throttle block output to the Multiply Const block input and the WX GUI Scope sink input. That's all we need to do to keep our CPU safe and sane. Before moving on, let's save our flowgraph as gain_db.grc.

The flowgraph should look like:

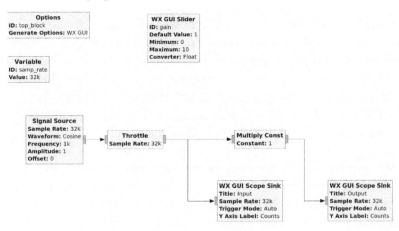

Now back to the decibels. To see what's going on with them, we're going to add some instruments that measure in dB and see what they tell us. First, we right-click on each of the WX GUI Scope Sink blocks, and select Disable. You'll notice that the blocks go gray when we do this, along with their connections. This feature allows us to make quick changes to our flowgraph that we can easily undo by re-enabling the blocks.

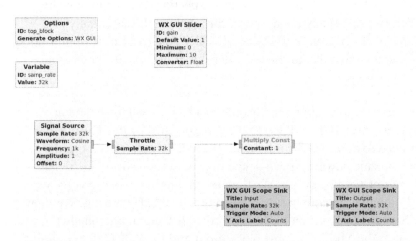

Then we add a WX GUI FFT Sink block, changing its Type property to **Float** and its Title to **FFT - Input**. Then we add a second WX GUI FFT Sink block, again changing its Type to **Float**, but this time its Title to **FFT - Output**. We connect these in place of the disabled Scope blocks, with the FFT - Input block to the Throttle block output, and the FFT - Output block to the Multiply Const block output. While we're at it, let's change the maximum value of our WX GUI Slider block to **100**.

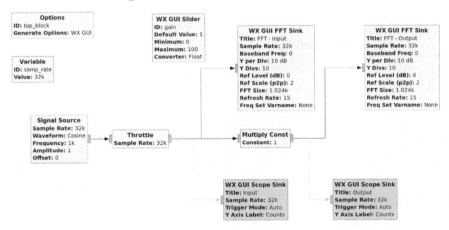

Now when we run the flowgraph, we can observe the effect of our gain property in dB. At first, the input and output FFT are the same, with a peak at -10 dB. We expect this, as multiplying any data by one should result in identical data.

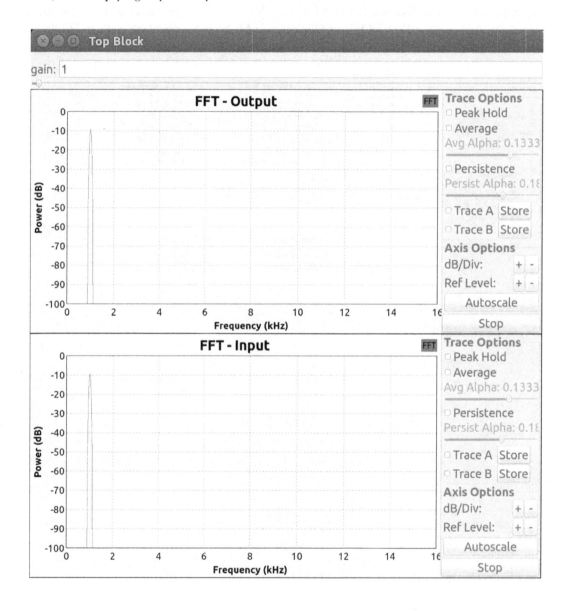

Then we change the gain to 10 (enter 10 and press Return) and examine what happens. You have to change the Ref Level to see the top of the output peak, but it's actually at +10 dB now. This makes for a change of 20 dB.

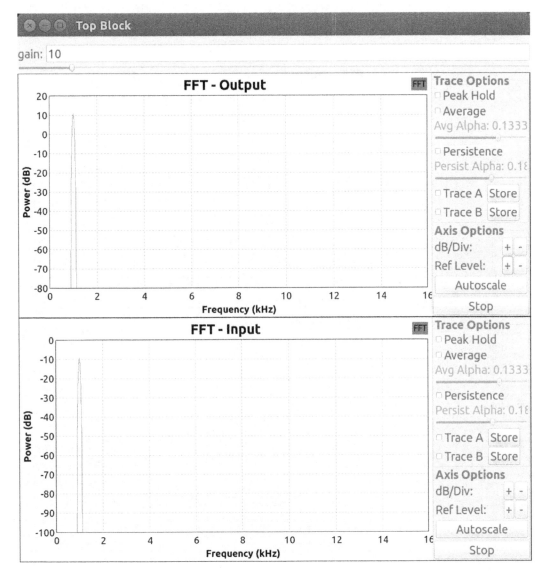

What do you think will happen when we change the gain to 100? What about 0.1?

The takeaway here is that increasing the signal's amplitude by a factor of 10 results in a change of 20 dB on the y-axis of our FFT. Put another way, an increase of 20 dB on our FFT means we have applied a 10x gain to the original signal.

Furthermore, subtracting 20 dB means a 1/10th attenuation of a signal, as you saw when you plugged in a gain value of 0.1.

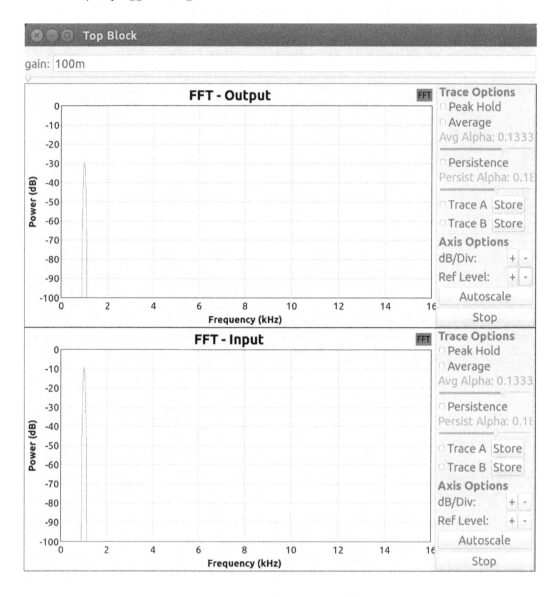

Before moving on, we do need to touch on one tricky aspect of working with decibels. Typically when working with real-world radio systems, a gain of 10x is equivalent to 10 dB, not 20 dB. So where did our 20 dB value come from? The key to this discrepancy can be seen in the y-axis label of our FFT: Power (dB).

When measuring gain it is most straightforward to compare apples to apples. In other words, if the amplitude of the signal changes by 10x, then this could also be expressed as a 10 dB amplitude change. In the specific case of our FFT, however, we are comparing apples to oranges. Signal amplitude is not the same thing as signal power, so the 10x change in amplitude is not the same as a 10 dB change in power. Without getting into the physics, just be aware that a 10x change in signal amplitude produces a 20 dB change in power.

The key takeaway is this: when working in gnuradio, you will most often see the 20 dB figure, but later on when you're working with radio hardware expect to see the 10 dB figure.

Despite this minor bit of confusion, decibels are a very convenient way of keeping track of the gain in a system. In fact, you can easily figure out the gain of a whole chain of blocks by simply adding the decibel-valued gains. For example, the same block diagram could be rendered in two different ways. First in simple linear gain notation:

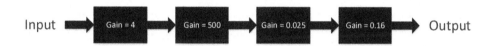

From this, we could figure the total system gain by multiplying each of the block gains or attenuations together. Pulling out our calculators, we see this mess resolves to a simple gain of 8 (where 4 * 500 * 0.025 * 0.16 = 8). If, however, we render the block diagram with gains in dB, we have the following.

Simply adding the decibel gains of the entire signal chain yields a total gain of 18 dB (where 12 dB + 54 dB - 32 dB - 16 dB = 18 dB). Recalling that a 10x increase in signal equates to a 20 dB change, you can see that this 18 dB value is just a bit less than 20 dB, just as 8x is a bit less than 10x. We won't go through the logarithmic math right now, but our answer passes the smell test.

Since most of the time in the radio world, gain will be given to you in dB, it's helpful to start thinking in these terms.

7.5 Filters

Now that you have a basic understanding of gain, it's time to dive into filters. Conceptually, you already know what a filter is. Think about a water filter for a moment. We pour impure water through it, and the filter stops (hopefully) most of the impurities from going through but allows the water to pass. On the far side of the filter we have our desired object: pure water.

In signal theory, filters have a very similar function. We feed a signal into a filter, which removes some parts of the signal we don't want, and a "purer" (or filtered) signal comes out the other side. Of course, when it comes to a signal, "pure" just means it has less of what you consider undesirable. What constitutes desirable and undesirable will vary depending on what you're trying to do.

Most of the filters you'll deal with are based on frequency. Earlier I talked about how signals can be broken down into their component frequencies via the Fourier transform. Taking this viewpoint, any given signal will be made up of a number of frequencies, some of which we might want and others which we may not.

Given that mindset, we want you to imagine what a low-pass filter will do. Hint: it has this name because it allows the lower frequency parts of input signals to pass through mostly unchanged, but stops most of the higher frequency parts of the signal from getting through.

Consider this FFT, with parts of the signal spread across both low and high frequencies.

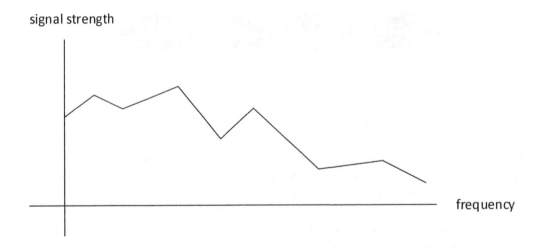

For the sake of this thought experiment, let's say that any frequencies above 2 kHz are "high" and any frequencies below that are "low". Note that there is nothing in physics or mathematics that defines this dichotomy, it is an arbitrary designation based wholly on what we, the filter designers, intend to do.

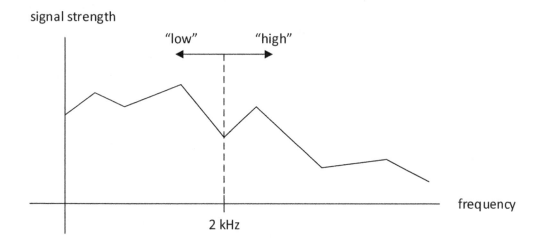

The goal of the low-pass filter is simply to reduce the part above 2 kHz as much as possible while affecting the part below 2 kHz as little as possible. In a perfect world, the filter would produce this:

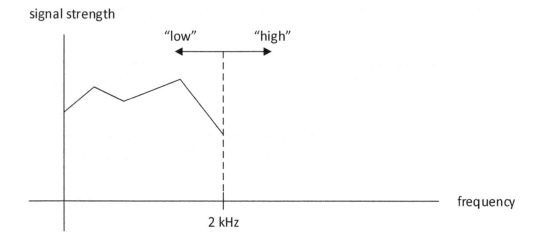

Think about this from an audio standpoint. You're listening to some music but are annoyed by a high-pitched humming sound. When you run your music signal into an FFT, as normal people do when troubleshooting their audio quality, what do you think you might see? Probably something like this:

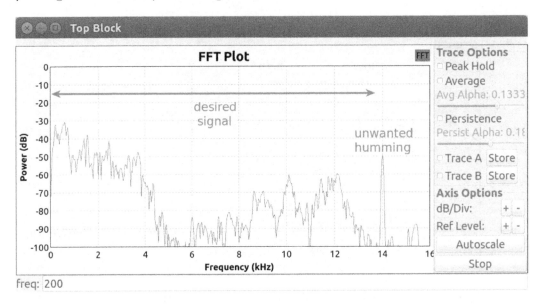

Let's say you've determined that the humming sound is about 14 kHz, while the music mostly ranges from 150 Hz to 12 kHz. Running the signal through a properly designed low pass filter should be able to eliminate the humming while leaving the music largely unaffected.

So enough with the conceptual. What do you need to know to use one of these filters? At our layer of the onion, just 3 things: filter type, cutoff frequency (or frequencies), and transition width.

Let's start by looking at the cutoff frequency using our cmajor.grc flowgraph from a few chapters back. We'll go ahead and rename this file to cmajor_lpf.grc. If you still have that flowgraph, you can save a bit of time by tweaking it to match what we're about to walk through. If you don't have it, no worries, I'll walk through it from scratch right now.

First, we place 3 Signal Source blocks, with Output Type **Float**, Amplitude equal to **0.1**, and frequencies (in Hertz) of **261.6**, **659.2**, and **1568**.

Then we place an Add block, changing its IO Type to **Float** and the Num Inputs to **3**.

Next we connect each of the Signal Source blocks to one of the inputs of the Add block.

Finally, we add an Audio Sink block and connect it to the Add block output. It's a good idea to execute the flowgraph now so we can verify it is set up correctly. You should hear a chord through your computer's speakers or from the headphone jack.

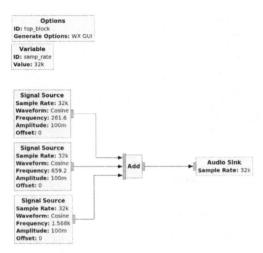

Now we are going to add a filter block to try to isolate one of the three tones. First, we'll go for the lowest tone. We start by breaking the connection between the Add block output and the Audio Sink by right-clicking and selecting Delete.

NOTE: You can also remove a connection by clicking and dragging it in any direction, as if you're pulling it off.

Next, we add a Low Pass Filter block and set its FIR Type to **Float->Float(Decimating)**. We also set its Cutoff Freq to **cutoff** and its Transition Width to **transition_width**. Then we connect the filter between the Add block and the Audio Sink block. Our completed flowgraph now consists of an interesting input signal, a filter, and a way to hear the effects of our filter.

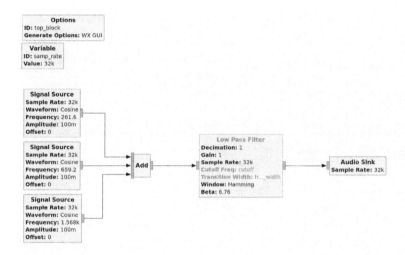

But as you can see, our flowgraph is still broken, as shown by the red text in the Low Pass Filter block.

To fix this situation we add a WX GUI Text block with ID of **cutoff** and a Default Value of **10e3**.

Then we add a second WX GUI Text box with ID of **transition_width** and a Default Value of **100**.

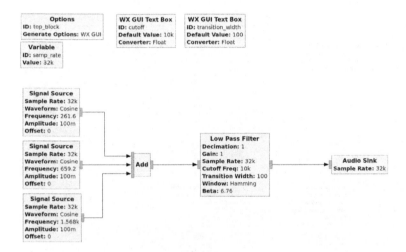

We could just listen to the output of our flowgraph, but there are better tools to analyze the frequencies in a signal than our ears. Hence, we add two WX GUI FFT Sink blocks and ensure the Type property of each is set to **Float**. We change the Title of the first one **Input** and connect it to the Add block output. Then we title the second FFT **Output** and connect it to the Low Pass Filter block output. Since we are using the WX GUIs, we need to make sure the Options block has the Generate Options set to **WX GUI**.

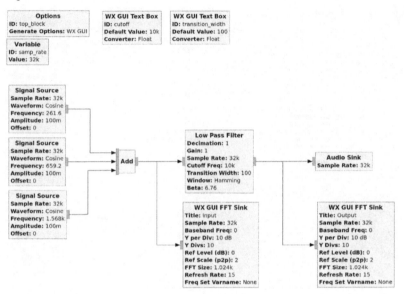

When we execute the flowgraph, our ears tell us that nothing has changed. The FFT displays also show that all three tones are still there.

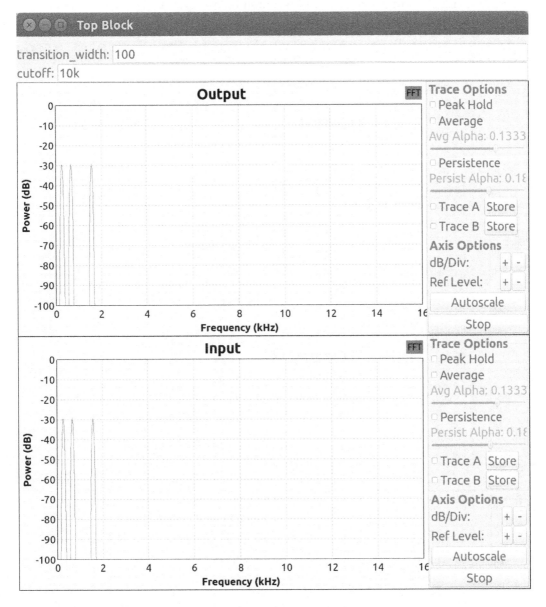

Perhaps this makes sense to you based on your suspected meaning of cutoff frequency. Let me give you a more formal definition right now. The cutoff is that frequency where the filtering takes effect. It's what the low pass filter uses to decide what's low, and will pass, and what is high, and shall not. The transition width is also involved in this, but more on that in a moment.

Since our first goal was to isolate the lowest frequency at 261.6 Hz, we need to set a cutoff higher than that. Also, since we want to make sure the second frequency at 659.2 Hz is filtered out, we need to set the cutoff below that. Accordingly, we change the cutoff to 300 and hit the Enter key.

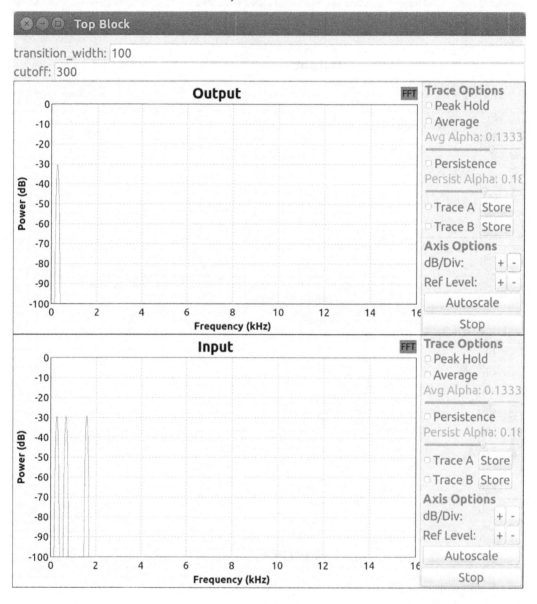

Hear the single low tone? Once again, you may need headphones to hear this. As expected, the FFT shows the low frequency peak is intact, while the other two are apparently eliminated. But are they really gone? To investigate we shift the output display downward by clicking the "-" button next to Ref Level on the Output FFT. After three clicks we see that both peaks are still there, just vastly diminished.

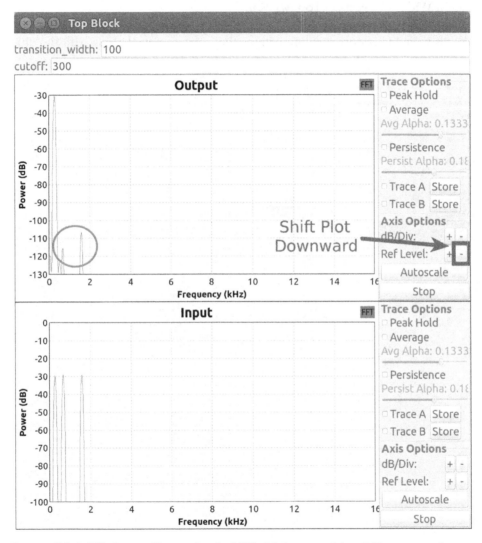

So we didn't kill them off completely. Will this be a problem? To answer that, we need to remember that the Y-axis of our FFT display is in dB. The other two peaks are about 80 dB lower than the first peak. Thinking back to our decibels section, each 20 dB represents a factor of 10, so our filter has reduced the unwanted frequencies by about 10*10*10*10, otherwise known as 10e4 or 10,000.

Before moving on, you should try to change the filter setting to pass the lower two peaks and filter the high frequency one. If you have trouble getting this to work, you can flip to the end of this chapter for the answer.

OK, we've figured out how to filter high frequencies while passing low ones. What if we want to do the opposite? For that, we use a high pass filter. Let's save the cmajor_lpf.grc flowgraph as cmajor_hpf.grc. To simulate the high pass filter, we delete the Low Pass Filter block and add a High Pass Filter block where we set the FIR Type to **Float->Float (decimating)**, set the Cutoff Freq to **cutoff**, and set Transition Width to **transition_width**.

We also need to restore the connections that were removed along with the old filter block.

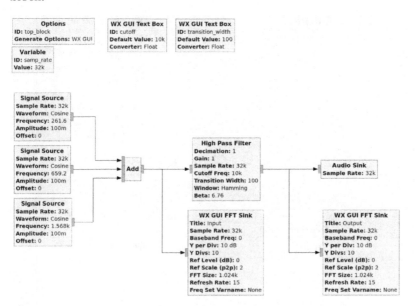

When we execute the flowgraph, we can see that all three peaks have been filtered. This is because we have our cutoff set at 10 kHz, meaning the filter considers anything greater than that a high frequency. Conversely it considers anything below 10 kHz to be a low frequency and is filtering, or eliminating it. This includes all three of notes in our C major chord.

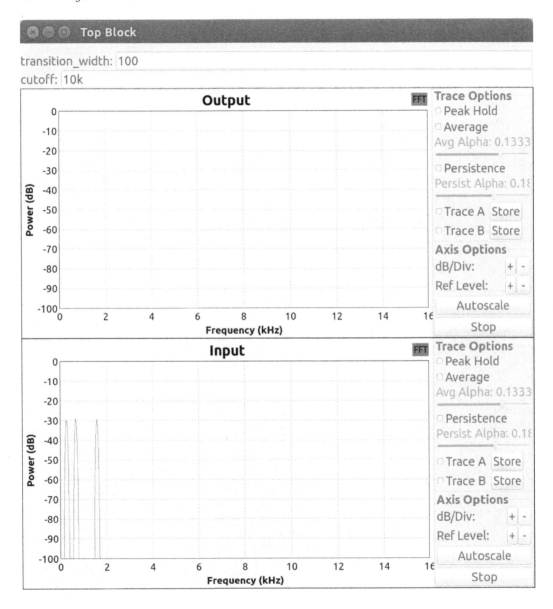

When we change the cutoff to 1300, however, we can hear a single, higher-pitched tone. The FFT also shows we have achieved the correct result.

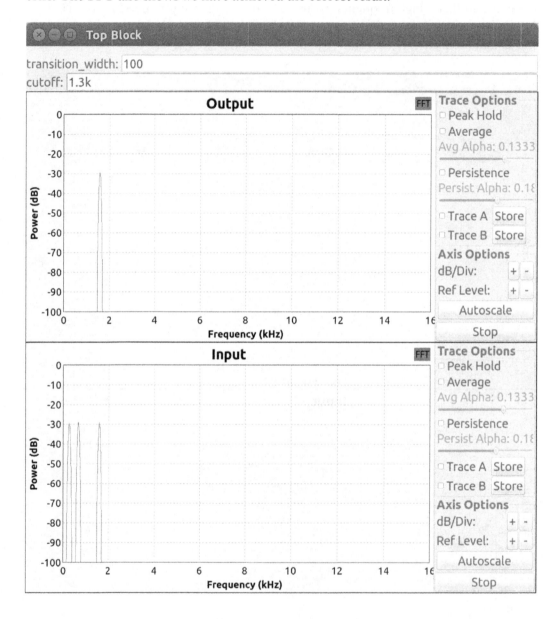

Here's a trickier proposition: how do we isolate the middle peak? Your first answer might be to use both high and low pass filters, which is a great answer and will definitely work. But there's already a filter type specifically designed to handle this situation: the band pass filter.

Let's go ahead and rename the cmajor_hpf.grc file as cmajor_bpf.grc. We start by replacing the High Pass Filter block in our flowgraph with the Band Pass Filter block, but notice something new in the properties window. There are two cutoff frequencies now! Showing no fear, we just label them **low_cutoff** and **high_cutoff**. We also change the FIR Type to **Float -> Float (Real Taps) (Decim)**, and use **transition_width** for the Transition Width property.

Because we now have two cutoff values, we take the existing WX GUI Text Box ID of cutoff and rename the ID as **high_cutoff**. We then add a new WX GUI Text Box, setting its ID to **low_cutoff** and its Default Value to **9e3**. (Note: whenever you use the Band Pass Filter block, make sure that the Low Cutoff Freq is always less than the High Cutoff Freq. This will prevent gnuradio from throwing execution errors.)

After repairing the broken connections we have this.

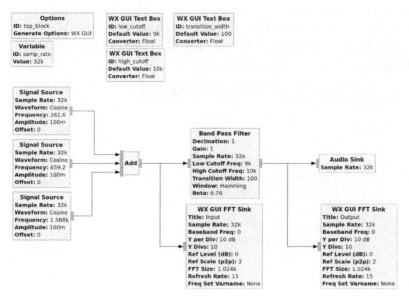

When we execute the flowgraph, however, we do not hear anything, nor do we see anything on our Output display. It turns out that we've filtered all three of the peaks again.

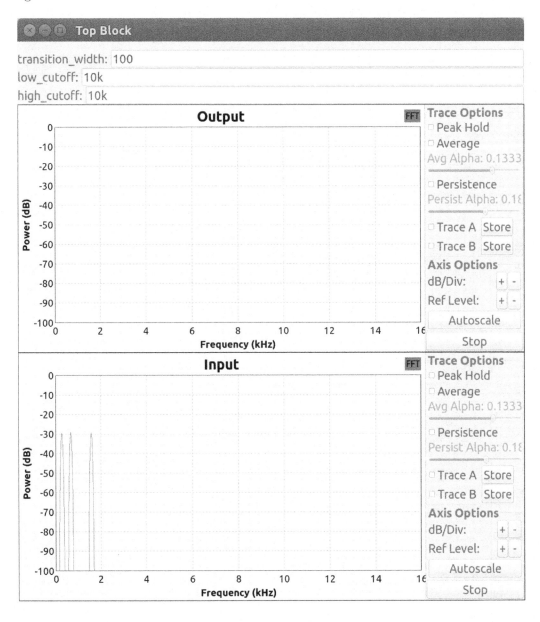

To fix this, we change the WX GUI Text Box value of low_cutoff to 300. Now we have set our filter's passband (those frequencies we desire to allow through the filter) to range from 300 Hz to 10 kHz, which contains the two right-most peaks. To shift the passband down to only cover the middle peak, we set the high cutoff value to **1300**. But looking at our FFT plots, something is still not quite right.

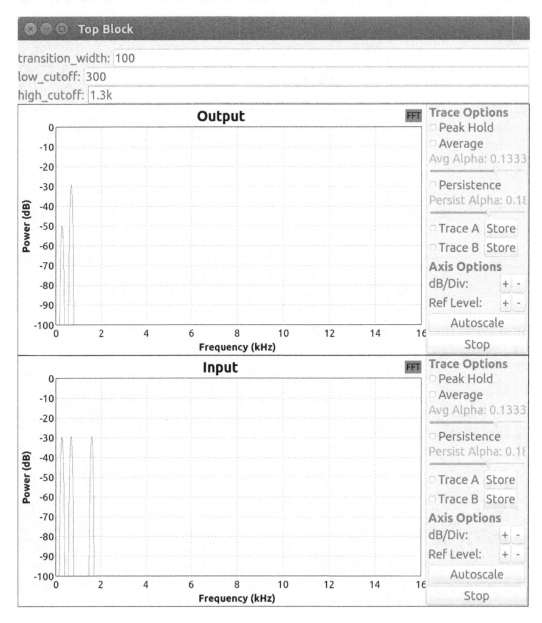

We seem to have only partially filtered the left-most peak. Any idea what happened here? To answer that, I should probably give you a long-delayed explanation of the transition width. The transition width is the range of frequencies between the cutoff frequency and the point where the filter is fully suppressing the unwanted frequencies. Here's a plot of the gain of a bandpass filter. Notice that in the transition region, some filtering occurs, but it is sub-optimal and worsens the closer you get to the cutoff frequency.

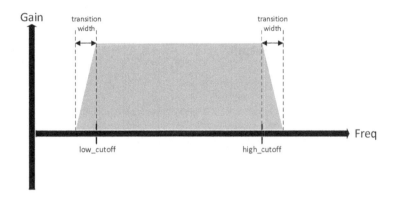

In our filter's case, the lower transition zone ranges 100 Hz from the cutoff frequency of 300 Hz, in other words between 200 Hz and 300 Hz. This means our 261.6 Hz lower tone has landed smack in the middle of the transition zone, resulting in partial filtering.

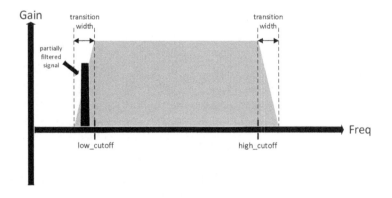

There are two ways we could fix this. We could move the low cutoff to the right, so it's more than 100 Hz away from our low peak. Or we can reduce the size of the transition width. Let's try the latter. We have set this to 100 Hz in our flowgraph, but our low cutoff (300 Hz) is closer than that to the lower peak (261.6 Hz). We instead choose 20 to narrow the transition width enough to kill off that pesky lower peak. After making this change, the transition zone now ranges from 280 Hz to 300 Hz.

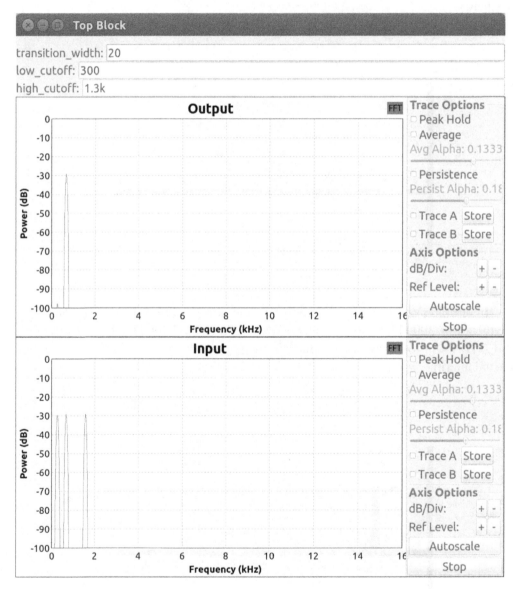

And there we have it. Only the middle peak remains. You will notice that the lower peak was not suppressed quite as much as the upper peak. This is because the lower peak is closer to the transition region of the filter. Don't worry, though, the leftmost peak is still more than 60 dB (or one thousand times) lower than the middle peak we wanted to pass.

One question you might have is "why not set the transition width to 0.0001 Hz and have an almost perfect filter?" The full answer has to do with digital signal processing algorithms, but the summary is that it takes too much computational power. The narrower the transition width, the more work the computer has to do to implement the filter. In general, it's a good idea not to make the transition width much more narrow than you need.

Before we move on from filtering, here's one final exercise. How would you suppress the middle peak while retaining the low and the high frequency peaks? Think for a minute and then check out the answer on the next page.

For the first filter exercise, you were asked a few pages back to configure the low pass filter to pass the lower two peaks while filtering the higher one. You can achieve this by setting the cutoff value of the Low Pass Filter block to a number between 659.2 Hz (the middle peak's frequency) and 1.568 kHz (the higher peak's frequency. A value of **1000** Hz works nicely.

For the second exercise (given on the previous page) you were asked to suppress the middle peak while retaining the low and high frequency peaks. To achieve this, you need a block called a Band Reject Filter (in the world outside gnuradio this may be called a Notch Filter). This block is the mirror image of the Band Pass Filter, possessing all of the same properties but with the opposite function. It suppresses, or attenuates, all frequencies between the low and high cutoff frequencies, while preserving those below the low cutoff and above the high cutoff. See the cmajor_brf.grc file for the solution here.

7.6 Equalizer Project

Before we go back to our AM radio receiver, we want to give you an optional project.

Think about your home audio system's equalizer. A typical equalizer will have a number of sliders that adjust the volume of your audio, but only at certain frequencies. For example, if you have the left-most slider pushed to maximum and all the rest to minimum, you will only hear the lowest frequencies of your music.

How would we go about building an equalizer in gnuradio? Assume three sliders, controlling the low-band, mid-band and high-band. Further assume that the low-band ranges from 20 Hz to 400 Hz, that the mid-band ranges from 400 Hz to 2600 Hz, and that the high-band ranges from 2600 Hz to 10 kHz.

If you get stuck trying to build this, you can take a look at the image below, or at the equalizer.grc flowgraph in your example folder.

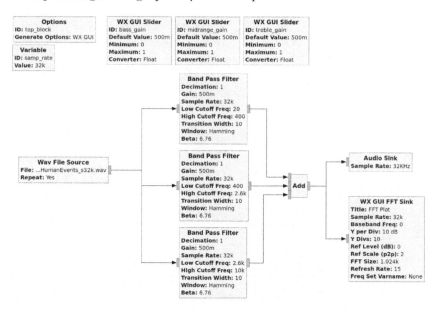

8   AM Radio Revisited

8.1 Memories

It is time to go back to the AM radio we put together earlier in this text. Here's a recap of what we've learned since then:

1) The idea of Amplitude Modulation (AM)
2) What signals are and how to think of their frequencies
3) What signals look like in the frequency domain
4) How to use Fast Fourier Transforms (FFTs)
5) The meaning of gain and how it is expressed in decibels
6) How and why to use filters

With these concepts under our belts, the radio should make a bit more sense now.

8.2 Input RF Data

First, we dig around on our hard drive for the first_am_rx.grc AM radio project we previously built and rename it second_am_rx.grc . Because we are about to connect a number of new instrument blocks in our flowgraph, we reorganize some of the blocks to make room.

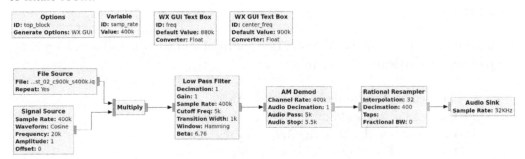

Let's start from the beginning, or the left side, of our flowgraph and follow the signals as they flow through to the right-side Audio Sink block. We first add a WX GUI FFT Sink block to the File Source block output.

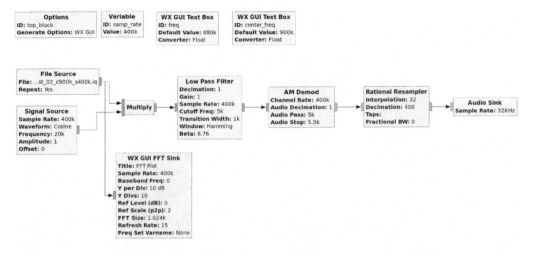

Note that the output tap on our File Source block is blue. Unlike many of the previous exercises, we don't need to change the Type of our FFT block to "Float." Interestingly, data coming from our radio file is complex. This is not simply a special case. In general, all radio data is complex. I'll get into why that is in Volume 2, but for now recognize that your radio data will start out complex and typically transition to a floating point format as you move closer to your sinks.

Our new FFT will allow us to see the frequencies contained in the radio data. Executing the flowgraph we take a look.

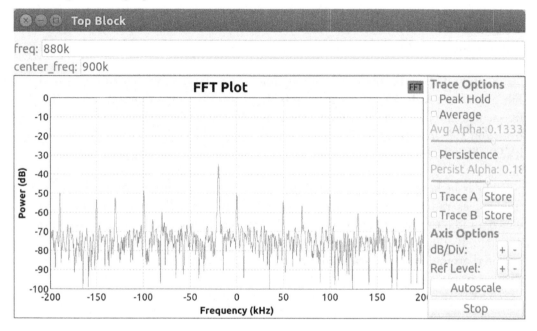

Our FFT data shows a few peaks occurring at suspiciously even values of frequency. This kind of spacing is what we'd expect to see in raw AM radio data: several stations transmitting and creating sharp peaks. We'd also expect those peaks to occur at multiples of 10 kHz, just like the AM radio stations which broadcast in your city (remember that the AM station number is just the frequency in kHz).

But there is something funny going on with the horizontal axis. Instead of ranging from 0 Hz to some number past 900 kHz, it's centered around zero. The values on the right side of zero make some sense, but what about the values to the left? And what on earth is a negative frequency? Well, I can't really explain how negative frequencies are generated at this layer of the onion, only that it's a consequence of dealing with complex numbers. But I can tell you what it means.

First off, I should explain something about the radio data we're getting from the File Source block. It doesn't range from 0 Hz to 900 kHz, as you might expect. The frequencies are much more limited than that. The file contains data only for the range from 700 kHz to 1100 kHz. This range might make a little more sense if I tell you that it's 900 kHz ± 200 kHz.

The issue here is that the original hardware captured a limited range of frequencies, rather than all frequencies starting from 0 Hz. The frequency range for which a radio receiver can grab data is called its input bandwidth. In this case, it was 400 kHz. By now you've probably noticed that the frequency range captured was centered around 900 kHz, based on the value assigned to the center_freq variable in the flowgraph.

But that still doesn't explain why the FFT is centered around zero, right? Well, it has to do with the nature of the sampled radio data. Even though it might seem as if the center frequency of the received radio signals would be embedded in the raw data, it's actually not. Again, this is for reasons I'll get to in the next Volume of our series. For now, realize that we need to provide a reference to the flowgraph so that it knows where our center frequency is. For the FFT, we can do that by changing its Baseband Freq property to **900e3** (900 kHz). An even better idea is to change the Baseband Freq property to **center_freq**, so the FFT will automatically adjust to any changes made to that value. In either case, this change will center the FFT display without changing the flowgraph's data in any way.

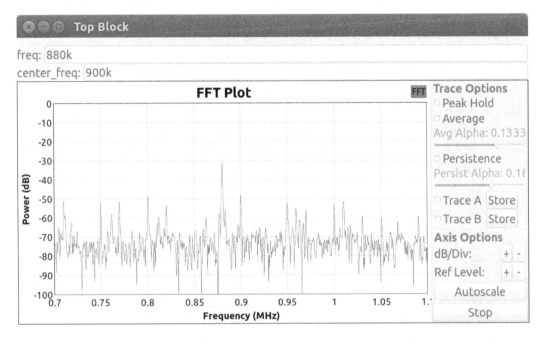

Now we have an accurate representation of our raw radio data relative to our horizontal axis. Notice how we have a peak at 880 kHz, which is the frequency we're currently listening to - based on the default value of the freq variable. Additionally, each one of the peaks in the FFT corresponds to a broadcast transmission to which we can tune.

Take a close look at this plot, as it is enormously informative and something we will be using in nearly every radio receiver that we design. It also vividly illustrates a portion of the radio frequency (RF) spectrum and how many signals share it. This is something I'll dive into much more deeply in the next section. Finally, thanks to our FFT, we're not limited to merely spinning the dial back and forth to find a station, we can just tune directly to the peaks we see. In fact, go ahead and tune to some of the other peaks by changing the freq variable. Very cool, right?

8.3 Tuning - Multiply

At the end of the last section, I asked you to tune your AM radio to some of the signals you found using your FFT display. But how does this flowgraph actually implement the tuning? The short answer is "multiplying by a sinusoid of the right frequency and then filtering." The long answer can be found in a digital signal processing (DSP) book. The in-between answer is contained in the next two sections of this chapter.

Rather than descend into the mathematics, we're going to focus on gnuradio experiments that demonstrate how this tuning works both in a simple project as well as our AM radio receiver. First, we're going to leave our AM radio and create a new simple project called freq_shift.grc. Note that gnuradio-companion can have multiple projects open at the same time, they just open in new tabs.

In our new project, we drop down two Signal Source blocks, one at 1 kHz, and the other at 2 kHz. Rather than change the Type to "Float" like we've done previously, we now leave them at their default **Complex** setting. As such, these signals will be more like the data we will get from actual SDR hardware. We'll also attach a Throttle block to the output of the 1k Hz Signal Source block.

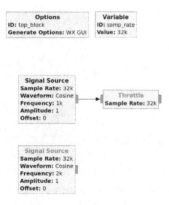

Then we'll add a Multiply block for the two sinusoids. The output of the Throttle block will go into one of the Multiply block inputs. The output of the 2 kHz Signal Source block will also feed into one of the Multiply block inputs.

Now place three WX GUI FFT Sink blocks into the flowgraphs to monitor each of the signals in the flowgraph.

First, we take one of the WX GUI FFT Sink blocks and attach it to the Throttle block output (to monitor the 1 kHz signal). We also set the Title property to **Input - Low**.

Then we take the second WX GUI FFT Sink block and attach it to the lower of the two Signal Source blocks (for the 2 kHz signal) and title it **Input - High**.

For the third WX GUI FFT Sink block, we attach it to the Multiply block output and title it **Output**.

Finally, we double-click the Options block and make sure that the Generate Options property is set to **WX GUI**.

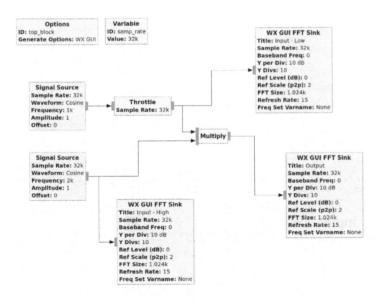

See what happens when we execute the flowgraph?

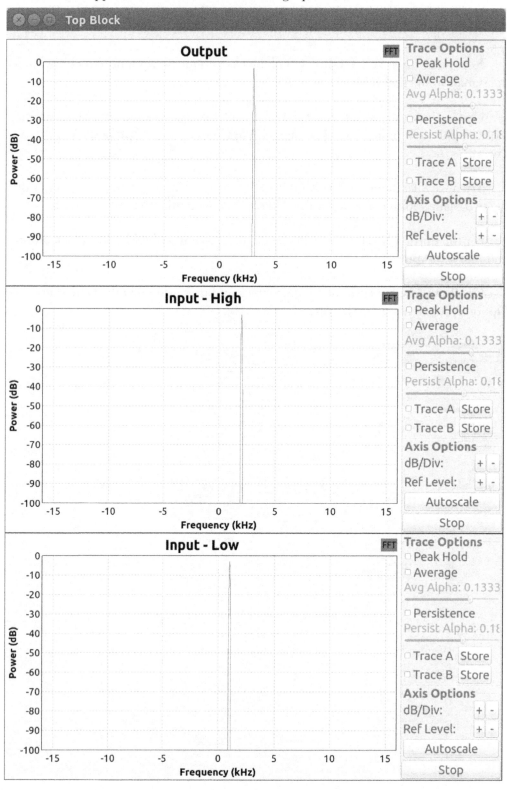

When two input sinusoids of 1 kHz and 2 kHz are multiplied together, a 3 kHz signal comes out. But is it a sinusoid? It looks like it is, but maybe we should check. We do that by adding a WX GUI Scope Sink block to the Multiply block output and disabling the Input-Low FFT and Input-High FFT so they don't take up so much screen space. To disable the blocks, we simply select them and press 'D' (or right-click the blocks and select **Disable**).

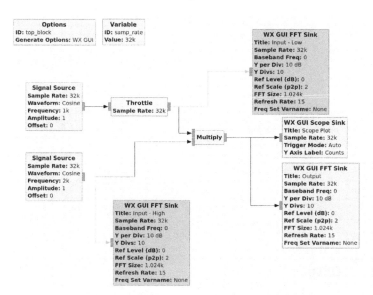

Now when we execute the flowgraph we can see our output signal in the time domain:

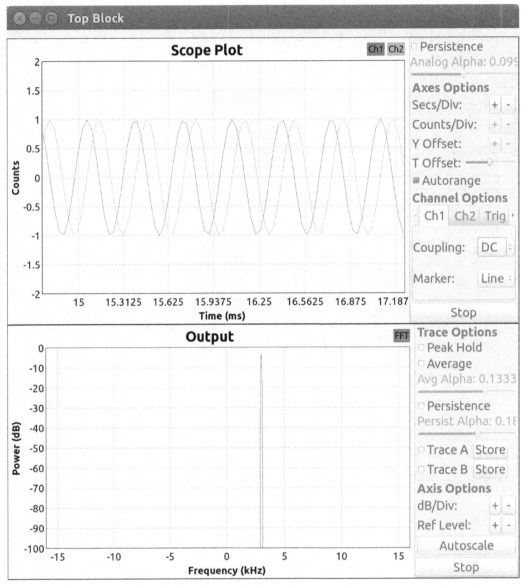

So it's a sinusoid! Or, more accurately, it's two sinusoids. Hold on, why are there two waveforms for a single signal? Well, that is the nature of complex numbers. You may remember back in math class (good times, eh?) that a complex number consisted of both a "real" and an "imaginary" part. The two waveforms you see are simply the real and imaginary portions of the complex signal measured. Let's not get too hung up on the word imaginary right now. We'll see later in Volume 2 (*Field Expedient SDR: Basic Analog Radio*) that the quantity called "imaginary" is a concrete thing that SDR hardware measures in real world signals. For now, just think of it as a piece of math terminology.

So we've generated some sinusoids. Big deal. What does that buy us? Let's try something else and see if the power of this technique becomes any more clear. We first rename the freq_shift.grc flowgraph as freq_shift_square.grc. The we disable the WX GUI Scope Sink block and enable the Input-Low FFT block and the Input-High FFT block. Finally, we'll change the Waveform property of our 1 kHz Signal Source block from Cosine to **Square**.

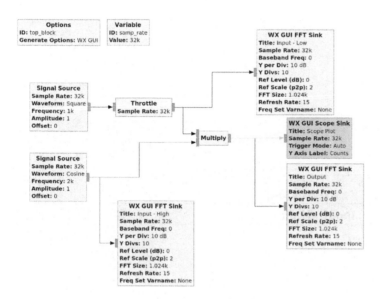

Why a square wave? Square waves happen to have the interesting characteristic of being composed of an infinite series of frequency peaks that stretch out in both positive and negative directions. Now is not the right time to delve into why square waves possess this characteristic, so let's just run the flowgraph to see what this looks like.

Looking at the Input-Low FFT output, we see the peaks continuing both to the right and to the left. The two biggest peaks, however, are at 0 Hz and 1 kHz.

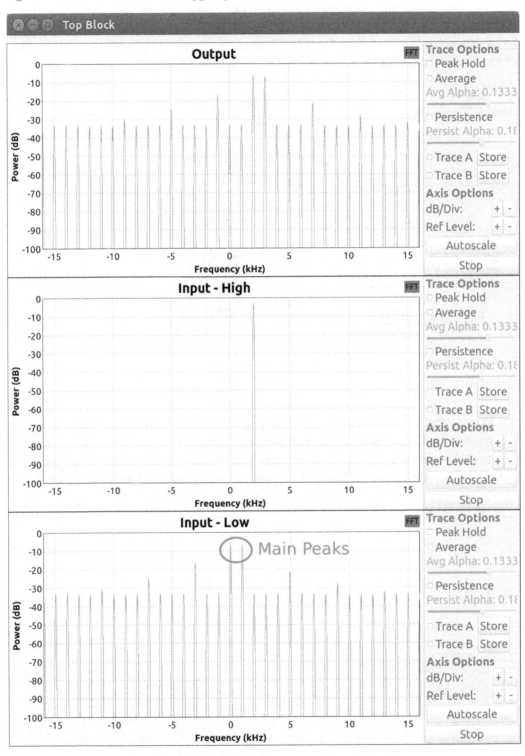

Now take a look at the FFT labeled Output. Can you see that the two largest peaks are now at 2 kHz and 3 kHz?

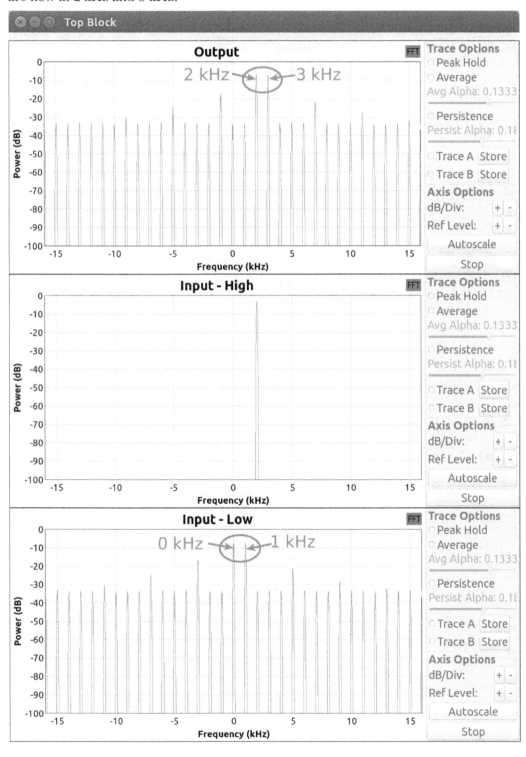

Comparing the locations of these big peaks on the Output FFT to those of the Input-Low FFT, we can see that they've shifted 2 kHz to the right. If you look closely, you can see that all of the other peaks have shifted along with the two big ones.

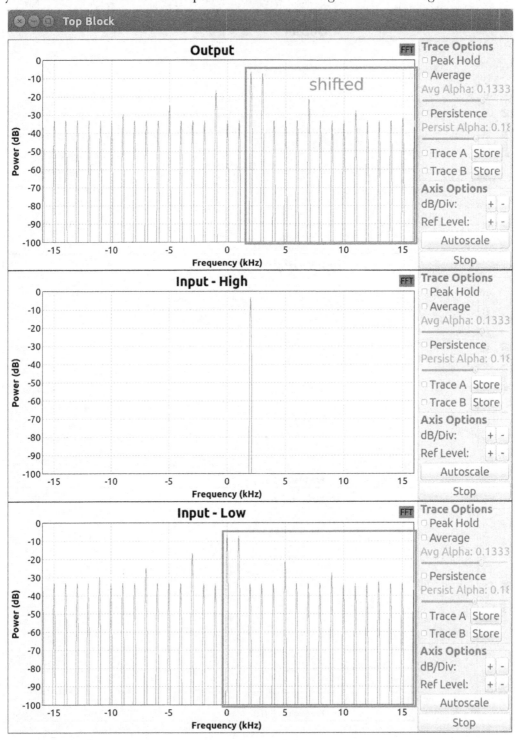

Congratulations! You've just demonstrated that you can shift the frequency of any signal simply by multiplying, in complex form, with a sinusoid. One thing that may not be apparent, is that you can shift both to the right and the left. To prove this we change the frequency of the second sinusoid to -2 kHz.

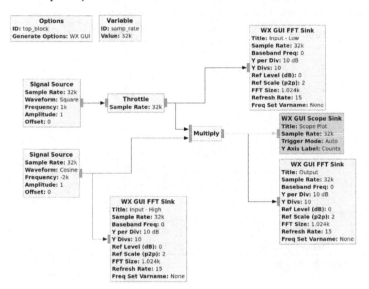

How can a sinusoid have a negative frequency? Again, it's a complex number thing we'll get to in Volume 2. For now, we'll just execute the flowgraph and observe that our two big peaks have now moved to the left by 2 kHz.

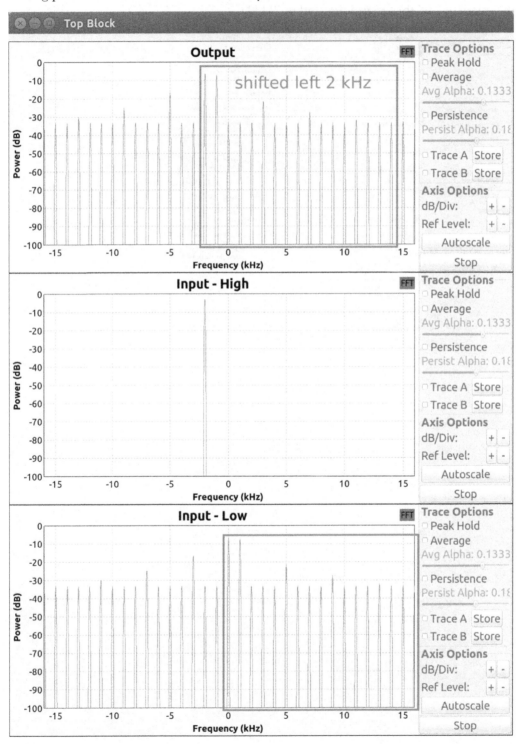

So let's go back to tuning for a moment. What does it mean to tune something? Well, in radio terms it means to grab only the signal at the frequency we're interested in, while ignoring all the other signals at other frequencies. In an old-school analog radio, this is accomplished using mixers and resonator circuits built with inductors and capacitors that filter out everything but the signals for which you are looking.

In the digital world of SDRs, it's a bit different. Imagine the data an SDR receives at any given moment. The FFT might look something like this:

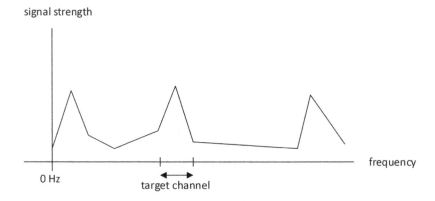

See how there are a number of different incoming signals, each represented by an FFT spike? To tune to the channel containing the target signal, we would first shift the frequency of our input radio data so that our target signal is at the center.

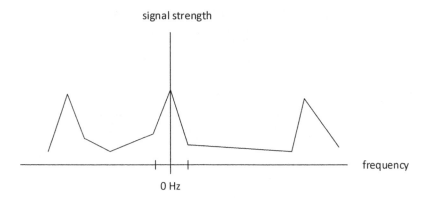

Then we would filter everything but the center signal.

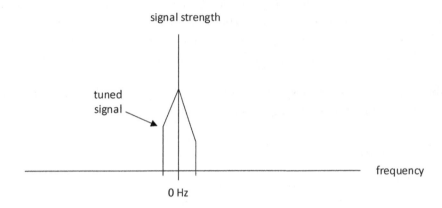

So we have two steps: shift and filter.

Let's see how this works in our AM radio flowgraph by opening second_am_rx.grc.

First, we copy/paste our WX GUI FFT Sink block and connect the input of the new FFT block to the output of the Multiply block.

Then we retitle the first FFT to **AM Input** and the second FFT to **Multiply Output**.

Since things on our flowgraph are getting crowded, you also might want to experiment with rotating the FFT Sink blocks. You can do so by clicking a block and selecting **Rotate Counterclockwise** or **Rotate Clockwise** from the Edit menu.

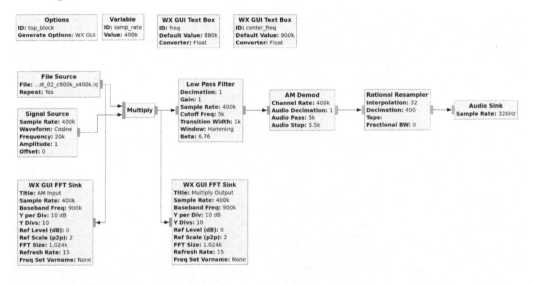

Can you predict what the new Multiply Output FFT Sink will show us?

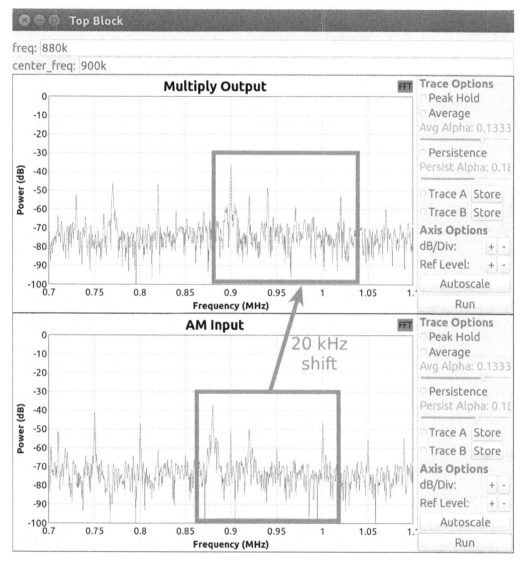

The input AM signals are all shifted to the right by 20 kHz. Since our Signal Source block has a frequency of 20 kHz, this shouldn't surprise us. Remember, the Frequency of the Signal Source block is center_freq - freq. This turns into:

900 kHz - 880 kHz = 20 kHz

As we change the freq variable in our flowgraph control (located just above the FFT displays), we can see the AM signals shift left and right. Let's try setting freq to **750k**. See how the FFT plot now shifts to the right by 150 kHz?

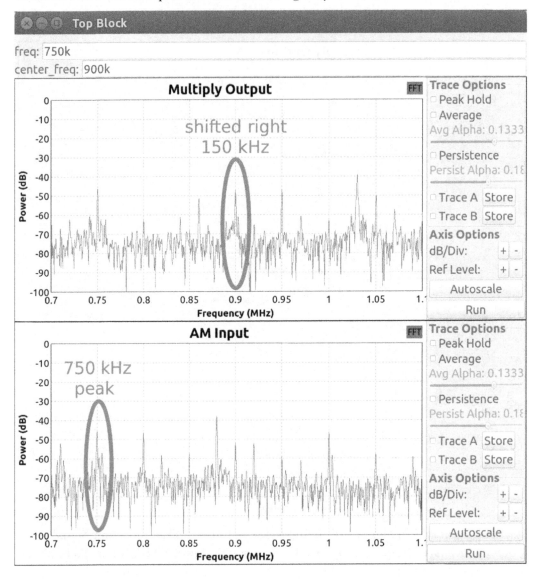

Here the frequency variable in the Signal Source block is 900 kHz - 750 kHz = 150 kHz.

So things are moving back and forth based on the value we enter for freq, which seems like the first part of tuning: shifting. But how is this working in the flowgraph? The key is understanding an expression we typed into our Signal Source block.

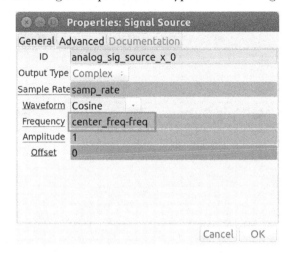

We previously set the Frequency of our Signal Source block equal to the expression **center_freq - freq**. This means that the radio data from our file will always be multiplied by a sinusoid with this frequency. Another way of thinking about this is that our radio data will always be shifted by a frequency equal to this expression.

As I told you in the last section, the center_freq variable is the center frequency at which the radio data was captured in the am_broadcast_02_c900k_s400k.iq file. When we look at the FFT of our AM input data, the middle of the plot corresponds to 900 kHz, our center_freq.

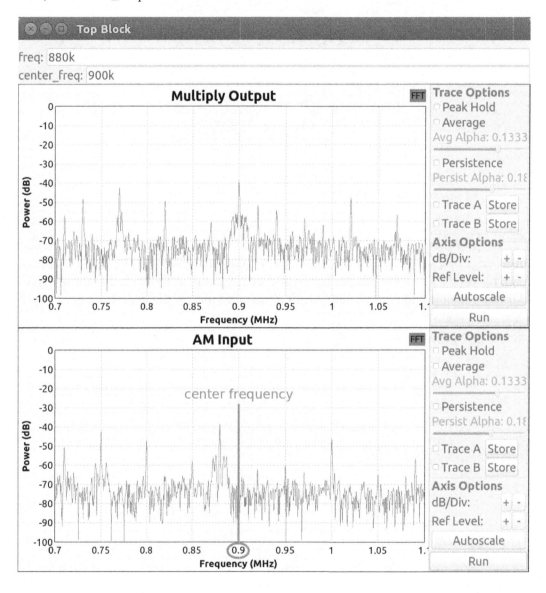

If we look at one of the many signal peaks, the one at 750 kHz for example, we can see that there is a difference of 150 kHz between the peak and the center frequency:

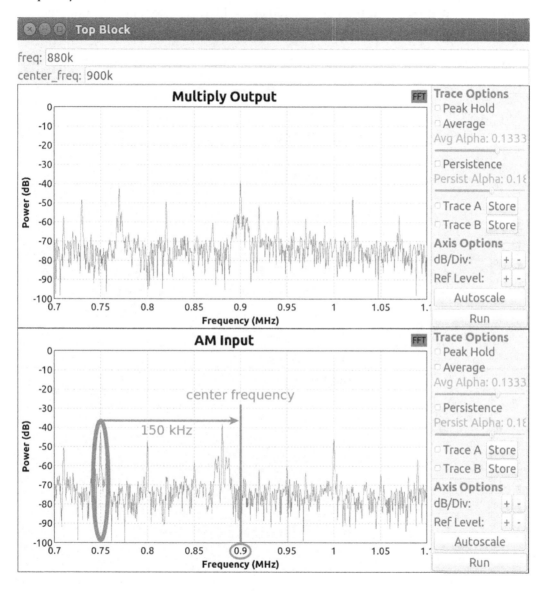

Assuming "freq" represents the frequency to which we want to tune, what expression should we use in the Signal Source block? In other words, how much do we want to frequency shift our radio data? I submit to you that "center_freq - freq" produces the desired result. If we look again at the FFT, any given peak is a distance equal to "center_freq - freq" away from the center:

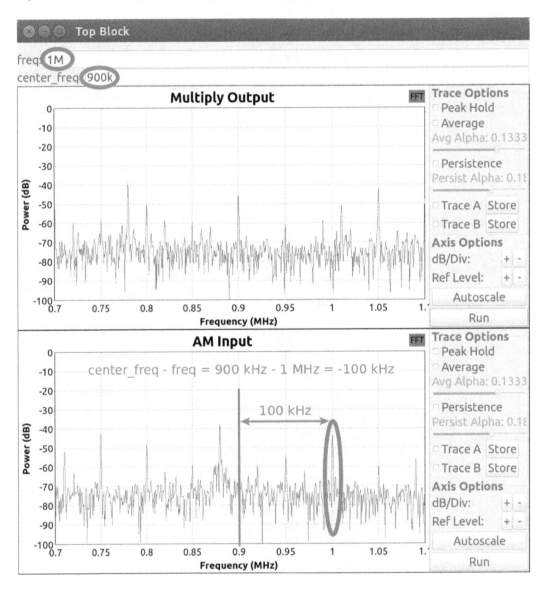

The "freq" variable is simply the frequency on which we want to center our raw AM data. As you can see in our Signal Source block, the frequency property is set to "center_freq - freq". Multiplying our input signal by this sinusoid will always center our AM data on the frequency to which we want to tune.

One last thing before we move past the Multiply block. Earlier, I mentioned that FFTs are all centered around 0 Hz by default, but that we can cause them to display their actual frequencies when we know them. For example, in this flowgraph, they are centered around 900 kHz because we've set that for the Baseband Freq property of each FFT. Don't get hung up on the meaning of the term baseband right now, that's a topic for a later volume. For now, let's change the Baseband Freq of the second FFT (Multiply Output) to **0** and rerun the flowgraph.

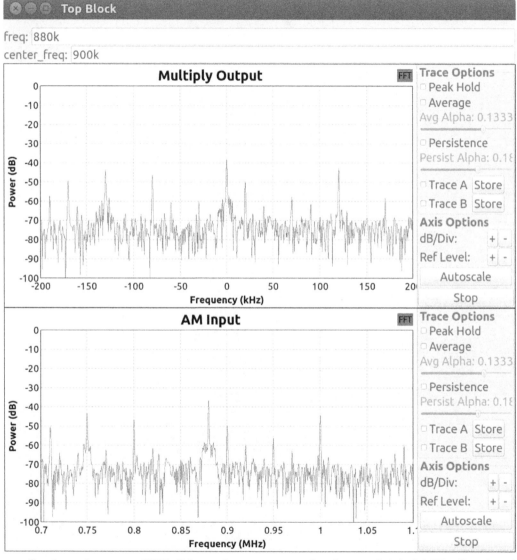

See how the upper FFT did not change, except for the x-axis labels? The Baseband Freq property does nothing more than change the numbers on the x-axis. After exiting the Multiply block, our AM data is centered not around 900 kHz or 880 kHz, but around zero. In fact, the input AM radio data is also centered around 0 Hz. This may all be a bit mind-boggling for now, but we'll revisit this idea in our next Volume entitled *Field Expedient SDR: Basic Analog Radio.*

Now, let's put all of the previous learning together and restate the purpose of the Multiply block. It takes the incoming radio signals and shifts their frequencies so that the one we want is now centered around zero. This is great, but there's still one problem. Per the definition of tuning above, we still need to grab only the signal we're interested in, not all of these others.

That's where the filter comes in.

8.4 Tuning - Filtering

To complete the tuning operation, we'll need to eliminate all of the other stuff that's not our original 880 kHz transmission. That brings us to our old friend the filter. Do you recall how filters can be used to eliminate frequencies we don't want while passing (or preserving) ones that we desire?

Thanks to the frequency-shifting Multiply block, we have the frequencies we want centered at 0 Hz. This should be good, right? We just need to pick a filter that passes the frequencies near zero and kills all the other ones. So what kind of filter would that be? Wait a minute!

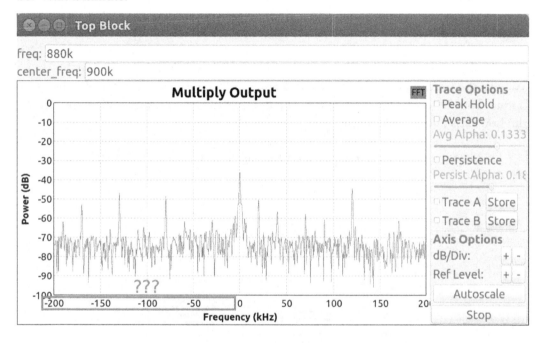

This is not the same as our previous filter exercises. We now have negative frequencies to deal with. What will a filter do with those? Before we can move on, we need to get a solid answer to that question. Rather than work through that answer in our complicated AM radio flowgraph, let's start a simpler project and see what happens. We'll call this new flowgraph complex_filter.grc.

First, we'll add a Signal Source block and set its Waveform property to **Square**. Remember how the FFT showed a square wave creating a series of peaks going in both positive and negative directions? This will give us an easy input for experimenting with filters and negative frequencies.

We next attach a Throttle block to the Signal Source block output to keep the computational load reasonable (remember when I talked about this?).

Next we add a Low Pass Filter block with a Cutoff Freq of **2e3** (aka 2 kHz) and a Transition Width of **100** (aka 100 Hz). As you might expect, we connect this block's input to our Throttle block output. We know this should eliminate all of the higher frequencies on the positive side of the axis, but who knows what they'll do to the negative ones?

Finally, we drop into the flowgraph two of our trusty WX GUI FFT Sink blocks. We take the first FFT, change the title to **Input** and connect it to the Throttle block output. For the second FFT, we change the title to **Filter Output** and connect it to the Low Pass Filter block output.

Finally, we remember to change the Options block Generate parameter to **WX GUI**.

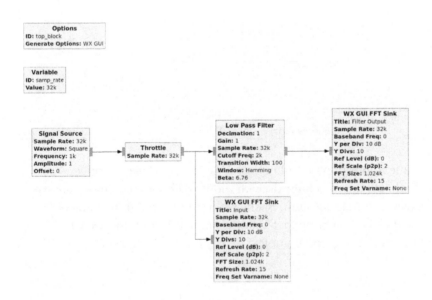

Holding our breath, we execute the flowgraph.

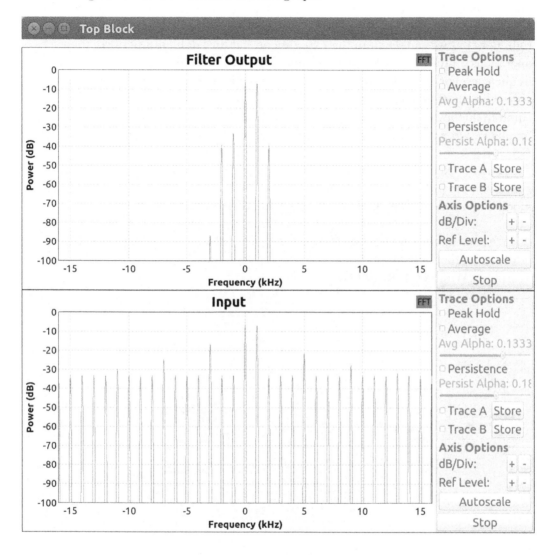

See what happened here? The filter attacked not only the high frequencies on the positive side of the x-axis, it also attacked the highly negative frequencies. Can you guess why this is the case?

It turns out that complex numbers are again the cause of an interesting phenomenon. You'll recall that our previous filter exercises used a "Float" type, which is just a way to express real numbers. We still can't dive too deep into complex numbers right now, but we can see from this experiment that they don't just perform their operations on the positive frequencies, but also on the absolute values of the negative frequencies. For example, the low pass filter will attenuate all positive frequencies above the cutoff frequency, as well as all negative frequencies below the negative cutoff frequency. Other basic filters behave the same way, operating on the negative frequencies in a mirror image of how they operate on positive frequencies.

A useful exercise would be to replace the low pass filter in our flowgraph with high pass and band pass filters to see how they each behave. Let's see how this works by modifying our flowgraph to use a high pass and then a band pass filter.

We start by deleting the Low Pass Filter block and adding the High Pass Filter block, using the same properties as before with the Cutoff Freq of **2e3** (2 kHz) and a Transition Width of **100** (100 Hz).

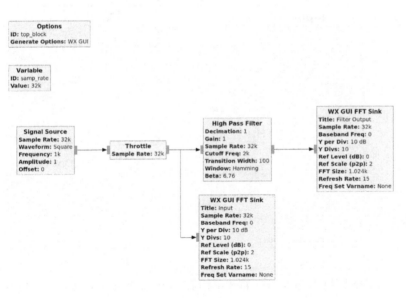

When we execute the flowgraph, we can see how it eliminates the lower frequencies, both positive and negative, while leaving the high frequencies untouched.

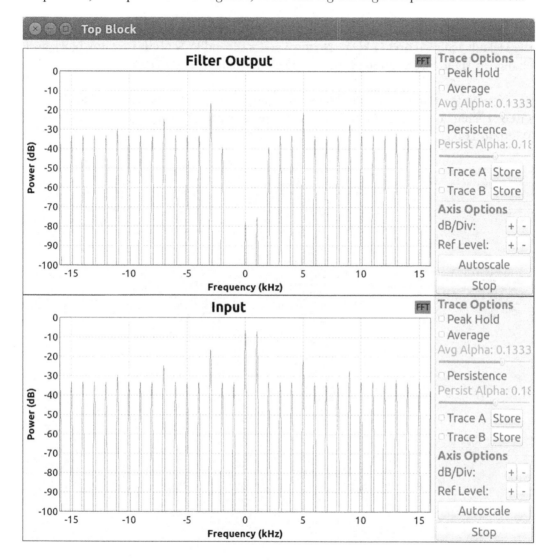

Next, we'll delete the High Pass Filter block and replace it with a Band Pass Filter block.

We then add a WX GUI Text Box block and change the ID to **low_cutoff** and the Default Value to **5e3** (5 kHz). Next we add a second WX GUI Text Box block and change its ID to **high_cutoff** and the Default Value to **10e3** (10 kHz).

Now, in the Band Pass Filter block, we change the Low Cutoff Freq to **low_cutoff**, change the High Cutoff Freq to **high_cutoff**, and change the Transition Width to **100**.

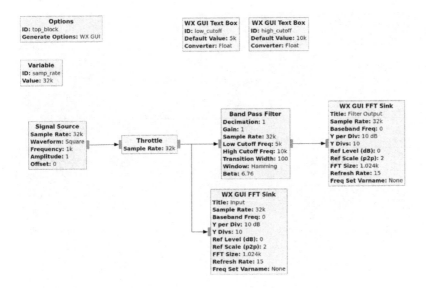

Executing the flowgraph, we see that the frequencies between our low and high cutoffs are preserved on both the positive and negative side of the plot. All the other frequencies are attenuated.

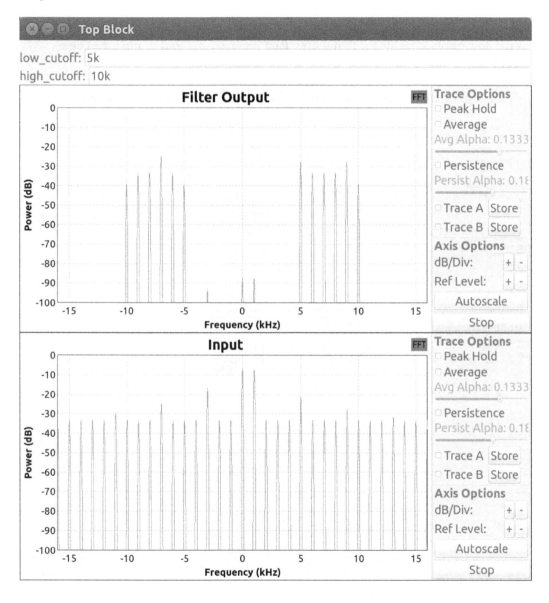

Moving back to our <u>second_am_rx.grc</u> AM radio flowgraph, we now have a better idea of what our filter block will do. Think about it for a moment and try to predict its behavior.

To actually show that behavior, we add yet another **WX GUI FFT Sink** block to the flowgraph, attaching it to the Low Pass Filter output and titling it `Filter Output`.

We also disable the AM Input FFT to keep the display clutter to a minimum.

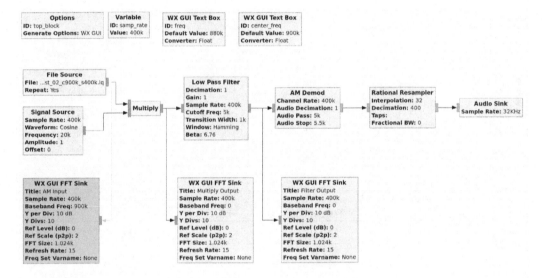

Running the flowgraph, we see the end result of our two-stage tuning process. (Remember that if you have trouble seeing the top of your FFT plot, you can always click the '+' button next to Ref Level to shift the display.)

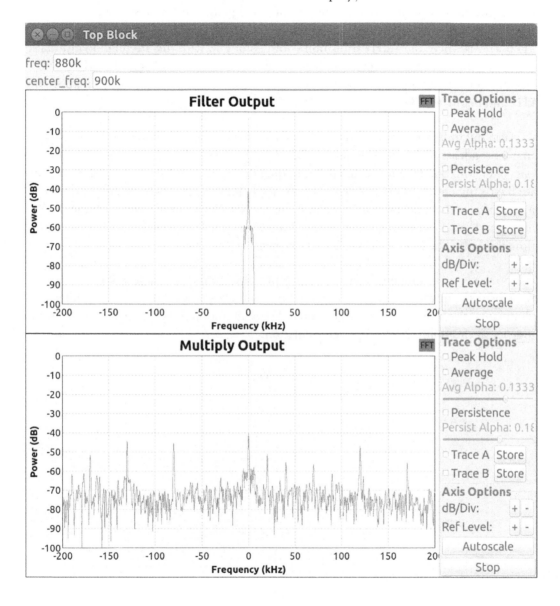

See how we have eliminated all of the other signals except for the one at the center of our FFT? Looking at the filter operation in more detail, we can see how the cutoff frequency of 5 kHz has allowed all of the input signal from roughly -5 kHz to +5 kHz to pass through, while attenuating the rest of the signals. You can also see that the signal does not "roll off" or slope downward in a perfectly vertical manner at +/- 5 kHz. This is where the 1 kHz transition width comes in, with the attenuation gradually becoming more severe between 5 kHz and 6 kHz, as well as -5 kHz and -6 kHz. Because the attenuation is not perfect, we can still see a few small peaks, even above 6 kHz and below -6 kHz.

You might be wondering why I chose 5 kHz for our cutoff frequency. It so happens that a broadcast AM radio signal has a bandwidth of 10 kHz. Since a complex low pass filter with 5 kHz cutoff will pass all frequencies from -5 kHz to +5 kHz, this means we'll pass the entire 10 kHz of signal bandwidth. This is a pretty brief description of a very important concept, but I'll go into this in more detail in the next Volume. I will also spend some more time digging into those filter properties we didn't talk about this time through. The filter onion has many layers.

So to recap, tuning has been a two step process in which we first shift our waveform so our target signal is centered around 0 Hz followed by a low pass filter that eliminates everything but our target signal. We're about halfway through our flowgraph journey at this point. Next stop: demodulation.

8.5 Demodulation

Remember earlier when I discussed amplitude modulation and demodulation? I provided a simple example where we modulated a faster carrier wave based on a slower sinusoidal signal. It looked something like this:

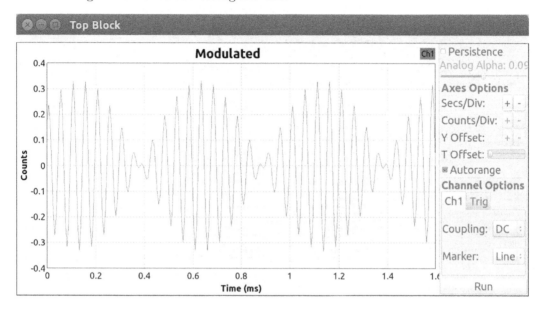

Now, coming out of our low pass filter, we have a single AM radio signal. It's been modulated in the same way as our simple example, but instead of a simple sinusoid, it contains an audio waveform. To demodulate this signal manually would require a fairly complex design, with scary things called "envelope detectors" or "product detectors." Fortunately, gnuradio has a simple block that just takes care of all that.

The AM Demod block takes a single AM modulated signal (which we have) and outputs an audio waveform (which we want). Looks like this is going to be a short section.

Taking a look again at our <u>second_am_rx.grc</u> flowgraph, we can see the complex AM signal going from the Low Pass Filter block into our AM Demod block. Then we see a float signal leaving the AM Demod block and going into the next stage, a Rational Resampler (whatever that is, more on this later).

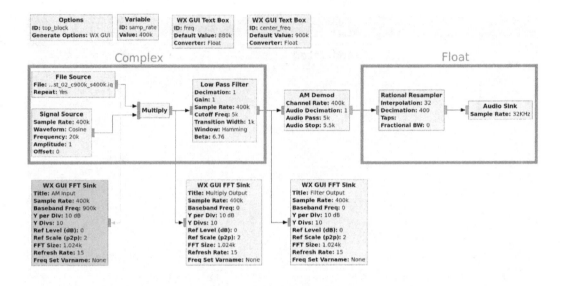

We know the data types in the flowgraph because of the blue tabs (complex) and the orange ones (float). As I mentioned earlier, radio data tends to be complex. I also mentioned that audio data will be composed of real numbers, float in this case. Often demodulators will serve as the dividing line between the radio side of things and the "rest-of-the-world output" side of things, which in this case is our audio.

Before we move on, let's at least look at the data going into and out of the demodulator block in the <u>second_am_rx.grc</u> flowgraph. We do so by dropping our favorite instrument block yet again: the WX GUI FFT Sink. We connect it to the AM Demod block output, title the FFT as **Demod Output**, and set the Type property to **Float**.

Before running the flowgraph, we also disable the Multiply Output FFT - we don't need it anymore.

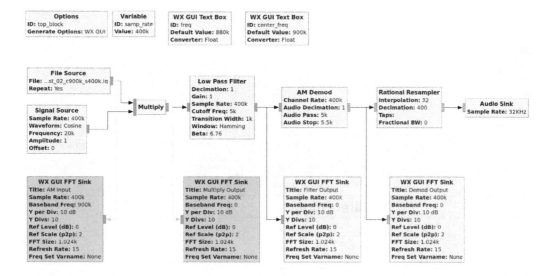

Think about what this new FFT block will show us when we run the flowgraph.

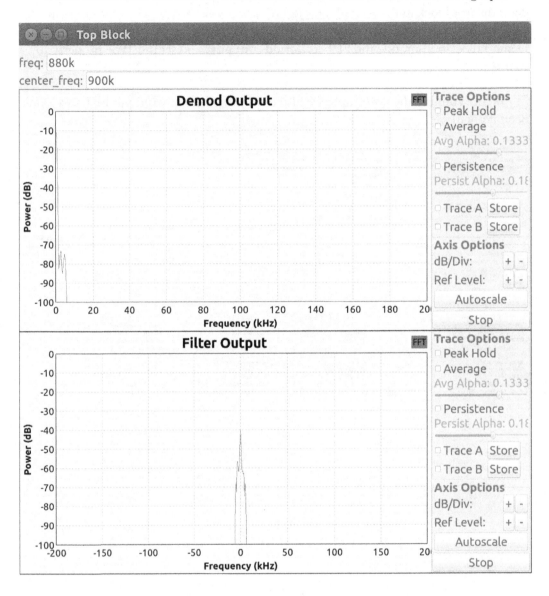

As you may have expected, the demodulator block takes in a signal with both positive and negative frequencies and outputs a single sided signal - one with only positive frequencies.

The last thing we want to mention are the properties for the AM Demod block.

1) The Channel Rate is simply another name for the sample rate.
2) As for Audio Decimation, it is just simple decimation, a concept we'll get to in the next section.
3) The Audio Pass is another term for cutoff frequency. This block actually contains its own low pass filter to get rid of any noise that may be part of the input or generated by the demodulation process.
4) The Audio Stop is another way of describing the Transition Width of the low pass filter. Instead of defining the size of the transition zone, the Audio Stop just defines where it ends. Arithmetically, the Audio Stop minus the Audio Pass is equal to the Transition Width.

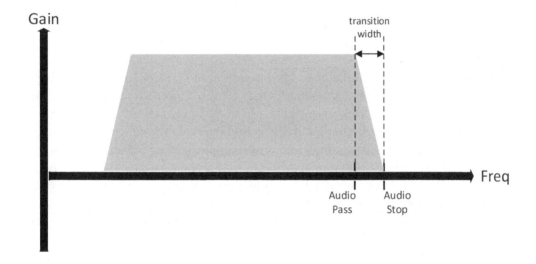

We'll get into more types of modulation and demodulation in later volumes. For now, let's be grateful for how powerful these gnuradio blocks are and move on to the last stage in our AM radio.

8.6 Resampling

So we tuned to our AM radio signal, demodulated it and produced an audio signal. We should be done, right? Just feed the audio signal to our sound card and that's it.

Well, it turns out there's one more snag (of course there is). You see, the sample rate of our audio signal is 400 kHz, as shown by the Channel Rate variable in the AM Demod block. But 400 kHz is too fast for the audio cards in most computers. This leads us to the first rule of sample rates in gnuradio:

Sample rates need to match when going from one block to another.

Put another way, the input sample rate to any block must be the same as the output sample rate of the block to which it's connected. So how do we change the sample rate? There are two ways: decimation and interpolation. Decimation reduces the sample rate, while interpolation increases it.

First let's talk about decimation, as it's the simpler of the two. Instead of thinking about rates, let's think about all the samples that are taken over a fixed period of time. For example, if our sample rate were 1 kHz (otherwise known as 1000 samples per second) we would have 1000 samples in a 1 second interval. To reduce the sample rate by two, we could simply throw away every other sample, leaving 500 over that same 1 second time interval. Since 500 samples divided by 1 second yields 500 samples per second, you can see that we've halved the sample rate for that 1 second period. Now rather than just doing this for a single second, imagine that we always throw away every other sample. If we do, then you can see how the sample rate would be halved.

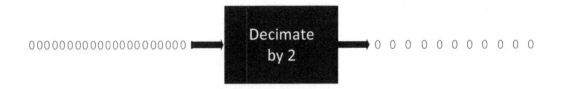

Let's create a simple project to see what this looks like in practice. Go ahead and save this project as <u>decimation.grc</u>. One note before we begin: we will be changing the Type on all blocks used in this project to **Float**.

We first add a Signal Source block and connect its output to a Throttle block. We will use the default values in these blocks

We then feed the Throttle block output into a Rational Resampler block, changing the Type to **Float->Float (Real Taps)** and the Decimation property to **4**.

Finally we attach WX GUI Scope Sink blocks to the output of the Throttle block and to the Rational Resampler block, titling them **32 kHz Input** and **Decimated Output** respectively. We also set the T Scale property to **0.5e-3** for both Scope Sinks. This will set the time scale of the x-axis display to 0.5 milliseconds per division, allowing us to easily compare the two plots. If we don't do this, the autoscale feature of the Scope Sinks will set them to different time scales.

As always, since we are using WX GUI blocks, we change the Options block Generate Options parameter to **WX GUI**.

The last thing we need to do is change the sample rate of the Decimated Output WX GUI Scope Sink from samp_rate to **samp_rate/4**. We do this because the decimator will reduce the sample rate, and if we don't adjust this property, the Scope Sink will not correctly display the data. Here's what the flowgraph looks like when it's done:

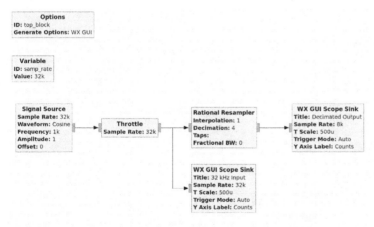

After flowgraph execution and the scope output windows come up, we change the Marker type on each plot to **Dot Large**. This allows us to see the individual samples rather than having a line connect them all together.

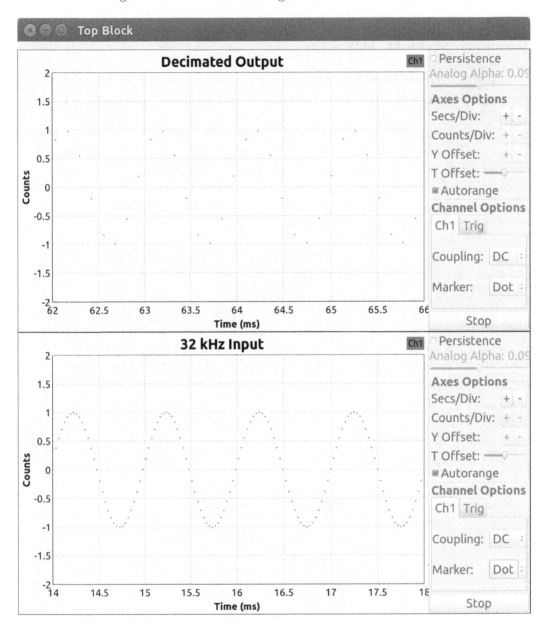

See how the Decimated Output waveform has the same shape as the 32 kHz Input waveform, but has fewer samples over the same time period? You can decimate any waveform this way, but be careful that you don't decimate too much or you will run afoul of the problem I mentioned earlier - distortion due to undersampling. I will give this issue more attention in the next volume of this series. For now, let me assert that the sampling rate in our simple test project is sufficient. After decimation by four, our sampling rate drops from 32 kHz to 8 kHz (32,000 / 4 = 8,000), which is still significantly more than the 1 kHz signal we are sampling.

One caveat: you can only decimate by an integer value. It makes sense to decimate by 7, for example, because we can keep every 7th sample and throw the rest away, reducing the effective sample rate by a factor of 7. The same method cannot be used to decimate by 7.5, for example. One cannot use the same method to keep 1 sample and throw away the next 6.5 without additional processing. Go ahead and try some different decimation values in the flowgraph to ensure it behaves as you expect.

OK, so you know how to reduce the sampling rate, what about increasing it? This too is possible through the interpolation process. As the name implies, interpolation creates extra samples between existing samples by estimating what value the original signal would have had at that point in time. For example, if our signal had a value of 3 at 1ms and 7 at 2ms, an interpolation algorithm might estimate the value at 1.5ms to be 5.

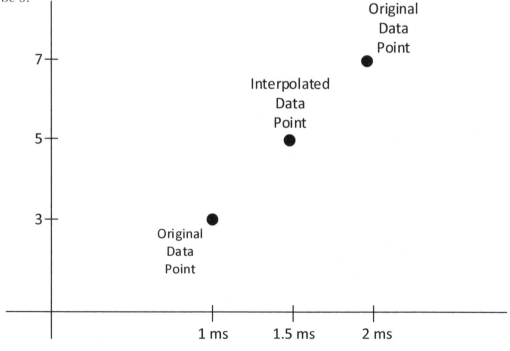

Let's change our flowgraph a bit to test this out. We'll save the decimation.grc flowgraph as interpolation.grc.

This time we reverse things by changing the Rational Resampler's Interpolation property to **4** and the Decimation property to **1**.

We then change the WX GUI Scope Sink's Title from Decimated Output to **Interpolated Output** and the Sample Rate to **samp_rate*4**. This is because we'll be speeding up the sample rate by a factor of 4 over the incoming 32 kHz.

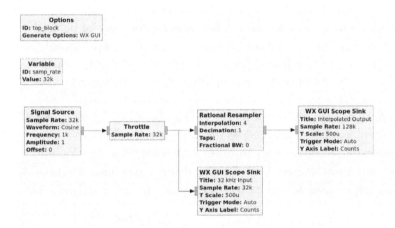

After executing the flowgraph we again change the Marker to **Dot Large** and we can clearly see the increased number of samples.

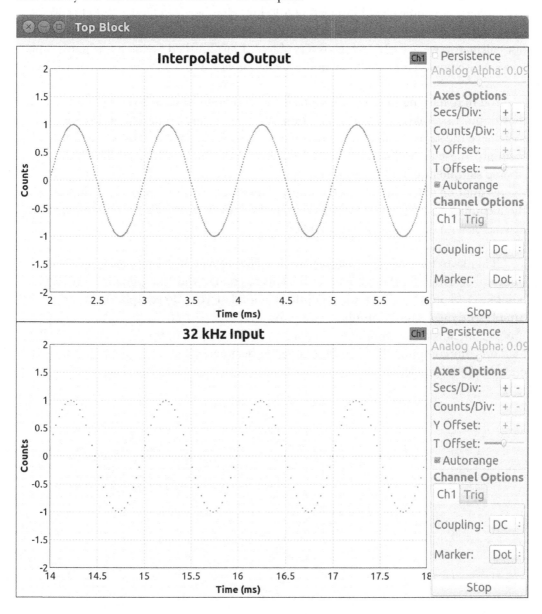

One warning on interpolation: we are not capturing any new information with interpolation. We haven't made any additional measurements of any real signals, we've just performed a mathematical trick that helps us match rates between our different flowgraph blocks.

Going back to our AM radio, let's see if we can get our sampling rates figured out. We're using a typical audio rate of 32 kHz for our sound card. At first it seems obvious that we have an audio signal sampled at 400 kHz that needs to feed a 32 kHz sink. So we decimate, right? By how much? Well, let's divide 400 by 32 and see. Punching into our calculator, we get 12.5, which is a problem. I just finished saying a few paragraphs ago how we couldn't decimate by non-integer values.

You may have figured out the solution to this conundrum by looking at the flowgraph. Essentially, we can do a combination of decimation and interpolation to achieve the same result as a fractional decimation. An easy way to do the math is simply to use the old sample rate in kHz for the decimation value and the newly desired sample rate in kHz for the interpolation value. In our case, we're using 400 for the decimation value and 32 for the interpolation values. This means that the rate of the incoming signal, sampled at 400 kHz, will be divided by 400 and multiplied by 32, leaving us with an output rate of 32 kHz!

Thanks to the rational resampler we are able to match our sampling rates in this project. However there are other reasons to change your sampling rate. A very common reason is to reduce the computation load of your flowgraph. A slower sample rate means fewer numbers going through that part of the flowgraph. Fewer numbers means fewer computations per second are required, thus reducing the CPU load on your computer. It's a good practice to decimate when you can for the sake of computational efficiency.

9 Conclusion

9.1 The Story Thus Far

So we've made it to the end of our introductory SDR journey. Well done! Before moving on to the next volume (or a celebratory beverage), let's recap what we've learned.

First, we talked about what SDRs are and the numerous ways they can be used to build radio solutions. We focused specifically on the ways that SDRs differ from traditional radios and what kinds of powerful capabilities result from those differences.

Next, we slogged through a bit of software installation. Hopefully it didn't take long. When you were done, though, you had an immensely powerful tool for working with radio signals, residing on your own computer. And the price was right.

I then gave you what must be the simplest description ever of how a radio works, with a particular focus on what it means to use one signal to modulate another.

Next, we tackled the issue of sampling. Specifically, we talked about how critical it is to sample signals fast enough if we want to capture a faithful representation of those signals.

We then turned our attention to gnuradio-companion and how to use it. You learned how to add blocks to your flowgraph as well as how to configure and connect these blocks. You also learned how flowgraphs pull in data using source blocks, perform operations on that data, and then output that processed data to sink blocks.

Next, we built an AM radio receiver without much understanding of what we were doing. Fortunately, if you're reading this, it means you had enough faith in your authors to persevere to the eventual explanation.

The frequency domain occupied the next several dozen pages. First, we learned what it means to visualize a signal in the frequency domain using FFTs. Then we practiced processing audio signals with the four basic filter types: low-pass, high-pass, band-pass, and band-reject. The frequency domain and filtering are probably the most important theoretical concepts we presented in this book. If they aren't clear to you, I recommend taking a moment to review that chapter before moving on to the next volume.

We then used our intuitive audio knowledge to talk about gain and decibels. Although I avoided the in-depth logarithmic computations associated with decibels, you should have enough of an understanding to handle them.

Finally, we went back through our AM radio flowgraph and saw how these concepts applied to the different stages of our radio. With each stage, we dug deeper to understand the specifics of how it worked.

In the first stage we introduced raw RF data into the flowgraph via a File Source block. We learned that the sample rate of the captured data is related to the size of the frequency range coming out of the source. If we had been building a real, hardware-based radio, this File Source block would have been replaced by a block that interfaced with our SDR hardware. No other changes would have been required.

Next our flowgraph tuned to the frequency of our target transmission using a combination of techniques. First, we shifted the frequency of the signal by using a Multiply block. Then we filtered the Multiply block output to eliminate signals other than our target.

We then demodulated the output of the tuning stage to produce our AM radio station's audio signal. Because the sample rates must be the same leaving one block and entering the next, we used a Rational Resampler block to match the rates between our AM Demod block and our output Audio Sink block.

Basically, we built a radio using your computer. In just a few minutes.

9.2 The Road Ahead

At this point, you should have a good handle on how to use gnuradio to build basic flowgraphs and debug them when there's a problem. Although you could jump off into the Internet and start figuring things out on your own, there's still quite of a bit of foundational radio and SDR understanding you need to build. The next two volumes in the Field Expedient SDR series will help construct that foundation and make you much more able to head out on your own into the SDR world.

The next volume in our series, *Field Expedient SDR: Basic Analog Radio*, shows you how to build real, hardware-based radios. This book also dives deeper into the radio theory we touched on here, fleshing out the concepts in more detail. Additionally, we show you what kinds of radio hardware you'll need in order to be successful with your SDR work. Throughout this second volume, you'll also learn to use progressively more powerful and complex gnuradio features.

Next, I hope you'll work through the third book in our series, *Field Expedient SDR: Basic Digital Communications*. This book changes the focus from analog to digital communications, diving into the different ways we can modulate digital data onto radio signals. We'll also explore how the data in digital transmissions is structured.

Both books are composed with a minimum of dry exposition and explain most concepts with practical, hands-on exercises using gnuradio-companion.

I hope that you have been pleased with gnuradio, both by its power and by how easy it is to use. Please consider continuing your SDR exploration with me as I proceed to bigger and better things throughout the Field Expedient SDR series.

Index

A

B

C

D

E

F

Made in the USA
Coppell, TX
19 December 2020

46351584R00116